AIRCRAFT DOPE AND FABRIC

—2nd Edition

AIRCRAFT DOPE AND FABRIC

—2nd Edition

by Ruth and Warren Spencer

MODERN AVIATION SERIES

TAB BOOKS Inc.

BLUE RIDGE SUMMIT, PA. 17214

SECOND EDITION

FIRST PRINTING

Printed in the United States of America

Library of Congress Cataloging in Publication Data

Spencer, Ruth.
Aircraft dope and fabric.

Includes index.
1. Airplanes—Fabrics. I. Spencer, Warren.
II. Title.
TL699.F2S67 1981 629.134'31 81-9180
ISBN 0-8306-2313-2 (pbk.) AACR2

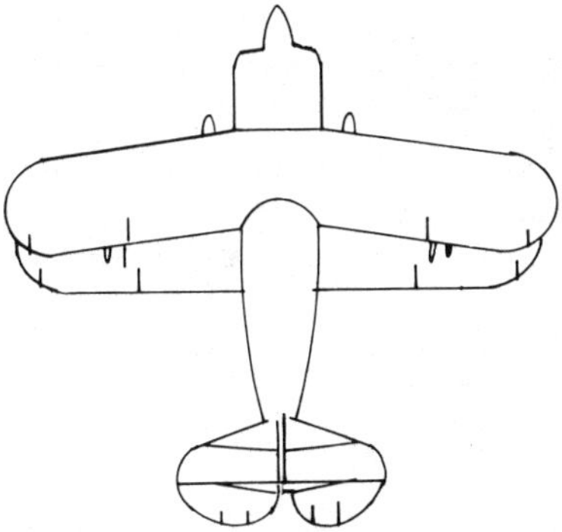

Contents

Preface 7

1 **A Backward Glance, Briefly** 9
Why Cover?—End Of An Era

2 **Preferred Covering Materials** 15
Traditional Materials vs. Synthetics—Flammability

3 **Materials, Tools, and Equipment** 21
Shopping Lists—A Place To Work—Other Essential Items

4 **Preparation For Covering Metal Surfaces** 26
Metalizing—Rust—Inspection—Drainage

5 **Preparation for Covering Wood Structures** 33
How It Shall Be—Wooden Wings—Removing Old Glue—Wood Glues

6 **Interior of the Cabin or Cockpit** 48
Why Cover Interiors?—Doing The Interior—Interior Trim

7 **Covering The Fuselage** 54
Measuring and Planning—Fitting the Material—Sewing—Gluing—Water Shrinking—Doping—Taping

8 **Wings and Tail Surfaces** 81
Measuring—Gluing—Shrinking and Doping—Stitching—Covering the Center Section—Miscellaneous Items—Patching and Re-doping

9 **Spraying Silver Dope** 105

10 **Spraying The Color** 114
Color Schemes—N Numbers—Blushing—Long Live Dope!

11 **Changes** 123
Rejuvenating—Enamels—Caveat Emptor—Synthetics—Pre-sewn Envelopes

12 **A Typical Recover With Ceconite and Pre-sewn Envelope** 132

13 **Appearance: Yes—Protection: Yes, Indeed!** 147
Wing Root Rot—Stits Poly Fiber—A Parting Word

Index 156

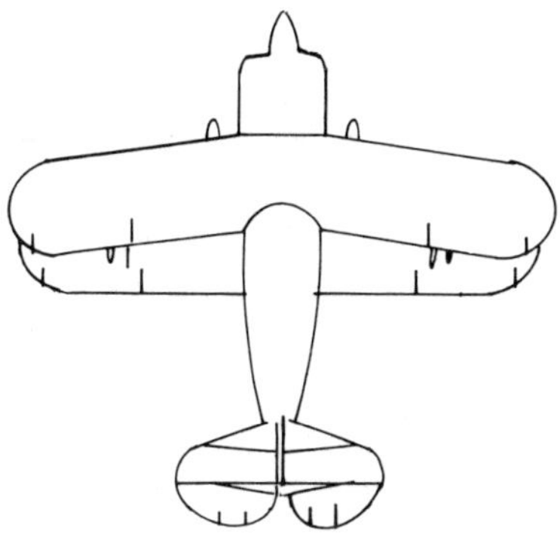

Preface

We wish to thank my son, Tom C. Street, for his cooperation in making possible most of the photos (of his Fleet biplane) used to illustrate the early chapters of this book, and his friend, Mr. Ben I. Hall, in whose home workshop the Fleet 16-B was overhauled and re-covered.

A favorite aunt once answered my comment about not having the opportunity to repay kindness and help extended to me, saying: "You do not *repay* it, you pass it along." So that is what we are trying to do for our new generation of aviators.

Photos are by the author except as noted.

Ruth and Warren Spencer

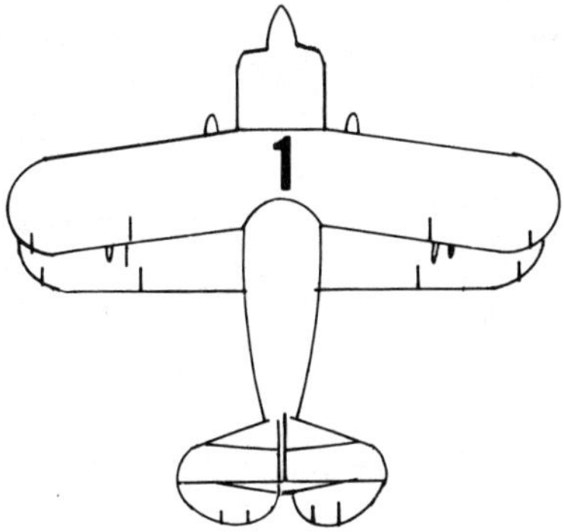

A Backward Glance, Briefly

The development of flying machines has been truly fascinating. Leonardo da Vinci lived between 1452-1519, and although many men before and many after him dreamed of and even built "flying machines" and attempted to fly in them, he took the first thoughtfully scientific approach. Many efforts were very imaginative indeed. But nearly four hundred years were to pass before man got off the ground in a heavier-than-air machine. Those who went aloft earlier in hot-air balloons were brave souls and true heroes. It is so easy to laugh at the early attempts, but when we stop and consider all that they did not have and had no way of knowing, all that they did do deserves our respect and gratitude. From the hot-air balloon to the moon in 70 years is almost beyond comprehension. Of course, for the present time, the moon is only for the few. Flying has always been only for the few relatively speaking. It rather seems that the next big problem is not getting off the earth, but finding suitable places to return to it.

Our concern here is with the airplane—not with balloons, helicopters or rocket machines. Here is Webster's definition of *Airplane*: "*n.* A fixed-wing aircraft, heavier than air, which is driven by a screw propeller or by a high velocity rearward jet and supported by the dynamic reaction of the air against its wings."—etc. And again Mr. Webster: *Plane* (pertaining to airplane): "One of the main supporting surfaces of an airplane or a wing,—;" and then the verb *Plane:* "To soar; specif., of a boat, to lift more or less out of the water while in motion, as a hydro-plane does."

So we see that unless a machine has "planes," or wings, it is not an airplane; it is something else. Without one or more wings the thing would never get off the ground—discounting rockets and the like—not within the range of horsepower commonly available to us today to provide the propulsion. That was where Leonardo was stymied some 400 years ago for lack of an appropriate power-plant. We are still looking for an ideal one. Maybe within another 70 years we will not need even the "planes" for flying. For the present we must use wings to fly our personal airplanes as we know them. They have improved since those earliest "flying machines."

The Greek myth of Daedalus and his son, Icarus, tells how they escaped a deep labyrinth (of Daedalus' own making) by flying up out of it with wings made of feathers. But Icarus flew too near the sun which melted the wax which held his feathered wings together and he fell into the sea and drowned. But quite obviously, the ancient Greeks studied the flight of birds and envied their ability to fly, as have all men from the beginning of time. The beauty, the grace, the freedom of birds has forever inspired men upward.

Maybe when we reach the outer limits of our earth's sea of life-giving air, we over-step our environment and such travel ceases to be *flying*; we ride then in projectiles, with propulsion but no wings. The outer atmosphere is too light and thin for the machine to be "supported by the dynamic reaction of the air against its wings." This we are familiar with as Pressure-Altitude, and woe to anyone who forgets it. We are making no complaint here; the newly developing aircraft are wonderful. But for the individual human being to own and fly and maintain himself, they are neither practical nor even desirable.

Let us return to our "planes" and their development; our substitute for feathered wings.

Why Cover?

One day a friend asked in all seriousness: "Why do you have to cover your little airplane? It looks so pretty like it is!" In my amazement I burbled some stupid answer about it being like people having to have skin over their bones—which has nothing to do with it and was no answer at all. I did a little better later when the shock wore off. This is not to belittle my friend; she is a wonderful cook and housekeeper and I am neither. She merely had observed from pictures that many of the very early flying machines did not have covering all over them. So here we go. . . .

The Wright brothers and their contemporaries were interested in *lifting planes*, sufficiently effective to get off the ground with the propulsion available at the time. At first, that was all they barely did. The engines were heavy and not very efficient. The structure had to be as light as possible. The wings of those first "aeroplanes" were of single camber only; there was no fabric on the underside. Ailerons came much bird-watching later.

Lifting efficiency improved noticeably with the application of the Bernoulli theorem to wing airfoils, utilizing differential in the top and bottom contours. There were a variety of airfoil forms tried before the beginning of the first World War, but many of the aeroplanes of that time had a minimum of covering other than on the wings and tail surfaces.

The De Havilland 2 (D. H. 2), one of Geoffrey de Havilland's earlier designs, was a popular and respected fighting plane in 1916. It was typical of the era. The full-forward cockpit and pusher engine nacelle was fabric covered. The aft structure which supported the tail group was a simple wire-braced open wood frame. At that period, the pusher engine, with cockpit forward of the wings, was favored for the excellent visibility and as a gun mount. Aerial battles called for a gunner to cover the rear. The engines were moved to the front with forward-firing guns up on top of the wing until synchronized "interruptors" were developed which very rapidly changed the whole picture. Model followed model as both sides competed for better performance in speed, altitude, maneuverability, firepower and durability. Engines gave the most trouble and they were the greatest procurement problem. One of the most obvious immediate changes when the engine moved forward was the covering of the complete fuselage structure. The drag of the open wood and wire framework of the earlier planes must have been terrific, although they were light in weight.

One wonders, though, if that covering really was much improvement. It was linen (so far, so good), but it was coated with banana-oil and varnish. Any old photo shows how rough, wrinkled and crude it was; the wings generally appear to have been better than the fuselages, fortunately. There does not seem to be much information about those covering materials. There are occasional brief notes of some experiments with other finishes. Transparent celluloid was tried on a B.E. 12 fuselage in 1916, and another in 1918 had an experimental pigmented "dope" and oil-varnish mix tried on its upper wing. Evidently, not much came of it. The 1918 Italian Ansaldo S.V.A. Scout was covered with clear-doped silk and it looked pretty good.

The various paint and color schemes used by both the Allies and the Germans are familiar to everyone, but the Germans' use of camouflaging lozenges *printed* into the cover fabric was the most unique and effective, being at the opposite end of the visibility spectrum from the gaudy paint jobs of the fighting planes of the most famous squadrons.

When we compare picture of the WWI aircraft and their crude finishes with our own today, we must realize those planes were turned out, from original design to flight test and acceptance, in only a few months. Working alone (which does make a difference), we take just as long to *cover* one airplane. But their only objective was to build a plane to carry a crew of one or two men, enough fuel for the mission and a camera and/or one or two guns with a couple hundred rounds of ammunition. They were not expected to last long and they didn't.

Our personal, more nostalgic link with the past, is the era of the 1920's and 1930's. Instead of the competition of war, the competition was in the fields of performance, capability and utility just to earn a place in the sun—to justify their existence. Speed records, altitude records, endurance records, new uses for airplanes and special-purpose types were developed. Lindbergh flew from New York to Paris in 1927. Admiral Richard E. Byrd and others explored the Antarctic by air. The ill-fated Dole Race saw two planes cross the Pacific to Hawaii. The National Air Races at Cleveland (most years) inspired a generation, including your authors.

Airplanes became faster, could carry bigger loads more efficiently and dependably, they were safer and more predictable, and they became increasingly expensive. Aeronautics became a science. But back in the twenties and thirties, the slow passage of an airplane overhead was something very unusual in the daily life of the average citizen; it was very special to a child who wanted to grow up to be an aviator. "Aviator" was then the commonly used term; not *pilot*. I think *aviator* means more than the word *pilot*. "To pilot" means just steering a boat or an airplane around from one place to another. Back when airplanes were very special, wonderful things, the aviator was the whole show. He not only flew his airplane, he nursed it and cursed it. He fed it and bathed it and fought for it, and he loved it. Sometimes he died with it.

Those were the days when an "airport kid" could learn first-hand how to take care of an airplane; how to dope fabric patches (nitrate dope was used then), clean dirty bellies and landing gears and greasy cowlings, watch for missing cotter keys and loose safety wire and worn spark plug leads and help swing a compass. He would go to the public library and look up books on everything he could find (pretty limited in small towns) on the history and theory and facts and fancy, anything related to flying. No one did it for him; no one, or practically no one, encouraged him. And if "he" was a girl, forget it! That was the time, during the 1930's in the time of the Great Depression, when the only way an FBO (another modern term) could afford to put gas in his own tank was to hop a few passengers on a good week-end; he was lucky if he had half a dozen students and maybe an equal number of planes based at his field to do maintenance on (he was usually *the* local A&E). A kid's first airplane ride at a cost of $2.00 was a never-to-be-forgotten event, almost on a par with first-solo. We both have first-hand knowledge of these things, and I'm happy to say that my son does too.

End Of An Era

By the time of the Second World War, airplanes were pretty generally accepted, but then they mostly went off to war. Hundreds of the fine old planes we loved were stored away in barns and old hangars and many of their owners never came back. The new ones *all* went to war; that was what they were built for. They were faster, bigger, more complicated, more costly and less appealing. Except for the Primary Trainers they were mostly all metal in construction; metal better lends itself to mass production methods. Meanwhile, back at the ranch, a lot of old biplanes went to work in a big new career as agricultural aircraft used for dusting, spraying and seeding crops and fighting fires in the timberlands. They are still doing it, hundreds of them.

After the war, in 1946, a new generation of private and business airplanes came on the market along with many surplus military aircraft. The new planes were efficient, but not very exciting. About the only biplanes in use were the ag ships, and the men who flew them and worked on them were a special breed. Ag pilots still are aviators, for with all the refinements in equipment and many added complications in the business end, the work remains basically the same and there is room here only for *aviators;* it is no job for dilettantes, or "heroes."

It has been only since the mid-1950s, when Robert L. Taylor rekindled interest in the "antique airplanes" and Paul Poberezny and a group of friends started the "experimental" home-built action, that the truly personal, fly for fun movement has snowballed. The national AAA, the *Antique Airplane Association*, with headquarters in Ottumwa, Iowa, has several thousand active members. Of course, there are not nearly enough old airplanes left to go around, but a sincere love of them will get you accepted in the group. Current membership in the *Experimental Aircraft Association* (EAA), based at Hales Corners, Wis., numbers over 50,000 at this time. Here too, the number of builders and owners is in the minority, but it is the interest that counts as far as the organization is concerned.

Both AAA and EAA stress quality, safety and pleasure, and that combination is hard to beat. In these days of space-age technology we are in some measure regaining a nearly lost value, and a lot of pilots are becoming "aviators."

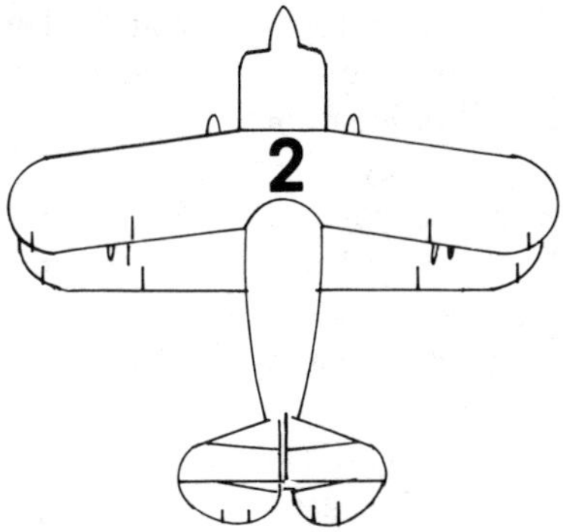

Preferred Covering Materials

One of the most annoying questions asked by the unknowing public is: "What's it covered with, canvas?" I mean to say now, does it *look* like canvas? The question seems at once stupid and insulting. You can only say (very politely) that it is Irish linen or a fine grade of cotton cloth (or whatever), not at all like canvas. I have a couple blouses which I made from Grade-A scraps after covering my J-3 Cub some years ago, and if I am wearing one of them when such a question is asked, I'll say, "See, the cover material on the plane is the same as this." When they see it in the natural state they are quite impressed with how nice it is, and they have learned something new. (*Wives, please note: Pre-shrink* and iron either linen or cotton *before* cutting to your size!)

We are *not* going to reprint charts or diagrams in this book, because anyone who intends to cover or re-cover any airplane should be familiar with, and have at his finger-tips, the FAA publication *Acceptable Methods, Techniques and Practices, Aircraft Inspection and Repairs* (with revisions) AC 43.13-1, available from the Government Printing Office, Washington D.C. 20402, or the older *Manual 18*, which is still applicable for dope and fabric work, and for most work required on the aircraft we are concerned with.

Whatever work is done on the ATC'd, Production Type Certificated aircraft and/or engine *must* be done according to standards published in the Manual. It is the licensed mechanic's "bible" and guide. Any work done in such a manner as to deviate from the approved methods is subject to rejection, which means replacing it properly. In all fairness, very little criticism can be directed against the Manuals. They have been evolved over long years of observation of failures and their causes, very often from faulty workmanship.

Whatever an owner does on his own airplane, beyond normal maintenance, *must be directly supervised* by an appropriately rated licensed mechanic. If you have your own *AC 43.13-1* or *Manual 18* right on the workbench and understand what it is telling you, you are a big step ahead in the right direction. We will refer you to it repeatedly.

If you are an "amateur-builder" you have a lot more leeway, but the Inspector will want to know precisely how you accomplished various procedures and he will feel better about the whole thing if you produce a well-worn Manual to accompany the working drawings you used. If he can see that you did do the work in an acceptable manner there will be very little static.

Any time you do not understand exactly how something is to be done, *ask someone well qualified*. Don't follow the "blind-leading-the-blind" system; that is just asking for trouble. But this is also a point that binds—just *who* to ask? You may well have decided to do this cover job yourself *because* you could not find any good shop or mechanic qualified or willing to do it for you—at any price. Ever since metal airplanes became their bread and butter, shops have turned down anything to do with major wood repairs and dope and fabric work. There has been a full generation of licensed A&P mechanics produced by FAA approved training schools with little beyond "this is wood and this is dope and fabric" included in their training. Anyone who really wants to learn about these things has to look up an old-timer someplace off the beaten path, one who runs a first class shop on the old type airplanes. They are scarce. There are a lot of old-timers around who are also strictly "shade-tree" operators who claim to know all the answers; it pays to *check credentials*. Some of these people don't know a good job from a poor one. So it pays to watch. Be observing; ask questions but don't believe all the answers you get; and learn for yourself how it should be done.

Our own experience is the accumulation of many years in widely scattered places and circumstances. We are approaching that dubious status of "senior citizens" in the ranks of airplane mechanics. We do realize how difficult it is to get advice on how to do a decent job, and this is the reason for doing this little book. We hope it will be helpful to both young licensed mechanics and to amateurs. With the increasing numbers of homebuilt aircraft, as well as the many restorations of older types, there promises to be many years of dope and fabric work ahead.

Since there will be repeated references to *AC 43.13-1*, do get a copy of this manual before starting any work and read the appropriate sections pertaining to the work you anticipate doing. It tells you *how it is to be*. The purpose of *Aircraft Dope And Fabric* is *to help you do it right*.

We suggest reading the book through entirely before starting to work, and then to refer to it as you work. Certain repetitions are not oversight, but are for emphasis.

Traditional Materials vs. Synthetics

At the outset, we state our personal preference for the Irish linen and Grade-A cotton fabrics, both used with cellulose acetate butyrate dopes—(for simplicity, called *butyrate* only)—as compared with any of the synthetic covering materials and finishes yet seen. We do not believe the newest thing on the market to be always the best. Nor do we consider all-metal aircraft to be in all ways superior to those of the older type of construction; it depends on the kind of flying you want to do and what you have need for. Fabric covered airplanes will be with us for a long time and we prefer them.

"Lifetime" durability is claimed for many synthetic covering materials but to date, none of them have been on any airplanes as long as some certain linen covered airplanes of our personal knowledge, so that point remains to be proven. We will continue to watch; the finish seems to present the biggest problem with synthetics. Some have been holding up very well and some not at all, although everyone said he had "followed directions" provided. We have felt that for our own planes we could not afford the risk of failure, nor have we been favorably impressed with these materials compared to linen or cotton for finish and appearance. This is our opinion and personal preference; we have seen some very nice cover jobs done with synthetics. Many people are experimenting and someone will come up with a fool-proof cover and finish someday.

With perhaps a few exceptions, synthetics are heavier than linen and linen is heavier than cotton. This is supposedly balanced by the need for less dope or special finishing material. But to achieve a finish comparable to linen or cotton requires *more* dope, thus increasing the weight. It seems that much of the apparent preference many people express for these materials is simply that they don't know how to use dope and fabric and they believe the ads that tell them it is "easier." Actually, except for shrinking and applying the dope or other finish, the procedures are practically the same.

Regarding durability, *no airplane should remain covered "forever."* Hangared and properly maintained, a linen and butyrate cover will outlast the time any airframe should go between overhauls—ten to fifteen years or even more, depending on climate. A first rate Grade-A butyrate job will go a good six to eight years with the same care. If an airplane must be kept outdoors both cover *and* airframe will need attention a lot sooner and it won't matter what kind of cover it has: wood will need at least sanding and revarnishing; 4130 tubing will develop rust and all aluminum structures do develop corrosion.

There is no shortcut to proper aircraft maintenance. Many advertised claims are sales pitches, no more. They don't lie; they just don't tell the whole truth. They want to see a product. Some are better than others.

The FAA does not say to use one thing or another, or not to use them. When a new product for aircraft use comes on the market its application is limited to specific aircraft on which it has been tried under supervision and has proved satisfactory. It is marketed under a special type certificate, on the terms that it will be applied and used according to an approved method, for each make and model of aircraft designated. So if you wish to use a new covering material and finish on the ATC'd aircraft, check first with your FAA General Aviation District Office (GADO) to be certain it has been —or can be—approved for use on your airplane. If you do use a synthetic, don't expect it to stay on "forever." The manufacturers or distributors must furnish instructions for use and they are best followed explicity.

The greatest advantage of finishes other than butyrate is that one is able to apply them regardless of weather conditions, and this can be an unquestionable asset when time is an important factor. I lost one full flying season because of persistent high humidity conditions.

Flammability

Regarding "newness" of materials, it is interesting to note that butyrate has not entirely superseded nitrate. Butyrate was a development of WW II and came into general use following the war. It is far superior to nitrate in every way, being more flexible, longer-lasting, and most important: its flammability is reduced. Butyrate does burn, but with an even progression, as compared to the instantaneous flash-up of nitrate. With a hot enough fire anything will burn, even metal; have you ever closely inspected a burned airplane? But nitrate ignites almost explosively and it takes very little to start it. The chemical relationship to nitro-cellulose is apparent in the name. Before WW II, when doping and sanding nitrate surfaces, they were always wet-sanded because of the fire hazard, and it was done outdoors if possible. There is no worry about friction-ignition or static electricity using the butyrate dopes. In a nitrate finished airplane, consider the possibilities: wiring problems with the ignition or electrical systems, static electricity, carburetor or fuel line fire, exhaust flames, etc. The lower initial cost of the nitrate dope just isn't worth it.

In regard to fire hazards relative to the fabric covering of any kind, we use aluminum over fabric in the area of the exhaust stack with any engine installation. Watch an airplane exhaust at night. The rate of burning of your plane's covering could mean the difference between getting out in an emergency, or not having a chance. To find out how anything burns, light a match to a small scrap.

Take the time to compare covering materials for the one best suited to your needs. We are not trying to sell you on anything. We prefer linen or cotton and the procedures given here are basically for their application with butyrate dope, though most details will be helpful in working with any material. Certain techniques apply to all and the elementary, basic procedures are common to all.

Irish linen exceeds all requirements for covering any aircraft and it outlasts Grade-A cotton about two-to-one under similar conditions and with an equivalently doped finish. It does take more dope to "fill" linen to get as smooth a finish as on cotton.

Grade-A cotton must test (tensile) new, at minimum 80 lb. strength, warp and fill. Aircraft fabric must have equal strength in both directions, which means equal warp and fill. Very often, linen fabric that has been on a plane for many years will still test higher than the 80 lb. new strength of cotton. For a light plane or a small one where weight is a very important factor, the lighter weight Grade-A is generally preferred. Hangared and given good care, it will last many years—as long as the airframe should really be allowed to go between overhauls.

An extra-light cotton fabric is approved for ultralight planes and gliders where only a few pounds count for a great deal. This is appropriately named *glider fabric*.

Note: If fabric is deteriorated, no amount of rejuvenation of the *dope* can possibly help the situation. The cover must be replaced. It is possible to rejuvenate the dope finish if the fabric is still good. Test a sample from the top surface of the wing. Rejuvenation of the dope is too expensive if the fabric is borderline and almost gone. In this situation, it is more practical to carry the fabric through just one more flying season by sanding and spraying it with enamel. This is the *only* time we recommend the use of enamel on any kind of fabric. It hardens and cracks and it is heavy; it is extremely difficult to patch even passably.

A good wax polish will give a butyrate finish a very nice slick surface and also protect it from water, dirt and oil. This makes it a lot easier to clean off the bugs.

3

Materials, Tools and Equipment

When an entire airplane is to be covered it usually pays to buy a full 100-yard roll of fabric. A light monoplane will require about 80 yards and a biplane, such as the Stearman, will use nearly the full 100 yards. There is no way of cutting to avoid considerable waste but it can be minimized by good planning.

Select fabric width which will best work out relative to wing panel length; the fewer pieces to be stitched together (edgewise) the better, up to a reasonable point. Measure the number of fabric widths required to overhang the total length (of the panel), deducting ½″ overlap for each adjoining seam to determine the width that will cover best without any seams laying over any ribs. One fabric width should cover the full side of the fuselage from top to bottom longerons; upper portions can be covered along with the top-cover piece. There will be unavoidable waste because very few components of any airplane are of regular shape—square or rectangular; they taper and have all kinds of weird angles and curves. We have found the 42″ width to be most practical for the average airplane. Extra-wide fabric is also very expensive and not really the most desirable for standard applications.

The leading merchandisers of covering materials can furnish quite accurate estimates on the amount of fabric, reinforcing tape, pinked tape, dope, thinner, etc., required for a standard airplane, along with a very helpful guidebook if you request it.

Shopping Lists

The following is a rough approximation for material for a light monoplane such as a Cub or Aeronca:

☐ 80 yards of fabric

☐ 5 rolls of pinked tape; 4 rolls of 2″ and 1 roll of 4″ (or 3″)

☐ 3 rolls of reinforcing tape (width of rib cap-strips)

☐ 1 roll of waxed rib-stitch cord

☐ 1 spool of machine-sewing thread (these are *big*)

☐ 1 spool of hand-sewing thread

☐ If screws and washers are used in your wings instead of rib-stitching, do not re-use old rusty ones; they aren't that expensive.

☐ Count all the necessary drain grommets for wings, tail surfaces, ailerons, flaps, fuselage, etc., and then get a dozen or so extra (see Chapter 8). They are very inexpensive and they do so much good. For an average wing, plan on 10 Inspection cover rings per wing panel, plus extras.

☐ One quart of lacquer cement (various trade-names are used) will serve for one airplane if it is not wasted. Once opened, it does thicken in time to the point of not being useable, but if it is quickly re-capped during the normal period of time of use it will last all right during the time required to cover the plane. It should not be gobbed on in large amounts but put on neatly *only* where it is to go, using a small brush.

☐ 30 gallons of dope (clear) and 20 gallons of thinner will be needed. Thinner is indispensable for many little jobs besides thinning dope, such as removing dirt, old paint, dope and varnish, etc. But it is fairly expensive so don't waste it.

☐ 2 pounds of silver paste will be adequate. We prefer this to the powder; it mixes better and is less "flakey."

☐ 4 gallons of pigmented dope (color), or if the area is divided about equally between two colors make it 2 gals. of each, should cover very well. Better to get more than enough than to run short when "just about finished." Pigmented dopes are almost always used straight from the can, so figure that what you buy is what will be used; don't count on its being stretched with added dope or thinner.

☐ 1 pound can of fungicide (anti-mildew) paste, which is mixed with the first coat of dope applied.

☐ 1 gallon of retarder—just in case it is needed.

Warning: Never buy dope or thinner advertised *cheap* in large quantities (or small). Nearly always, it has been bought at surplus for re-sale, already over-aged and deteriorated. Over-aged butyrate dope and thinner are *no good.*

Dope and thinner must be stored at a moderate temperature; never at or below freezing, nor at excessively high temperatures. Keep away from fire or flame in tightly closed containers. We had probably better add: "Keep away from glue-sniffers."

Synthetic fabrics have their own matching accessory materials: tapes, rib-stitch cord, thread, etc.

A Place to Work

The very first consideration is not *what* you need to do the work but *where* it is to be done, and for anyone without a workshop, this presents a major problem. A double, or even single, car garage is fine if the weather is warm (72° or above) and if there is adequate light and ventilation. If you have to work with the doors and windows closed, there would be neither. Where lights, and later on a compressor, are to be used, there must also be an electrical source in the building. I found out the hard way that there is too much current loss in a fifty foot drop-cord to operate even a small paint-spray compressor, and my plans to spray silver dope in the garage were modified and I transferred to the hangar for all the spray-doping. All of the hand-brushed clear-doping was done at home, then everything was hauled to the airport to finish. This may be just as well, because no matter how careful a person is in loading, there are always a few little scruffs resulting from hauling.

We do not recommend spraying dope, paint, or anything else in the basement (other than the little pressure cans), unless:

1) No heater-flame could possibly ignite the vaporized dope.

2) It can be closed off from the rest of the house reasonably. If a corner of the basement can be closed off, this would take care of 1) and 2).

3) There is ventilation, such as a window exhaust fan to remove excess dope vapors.

4) Temperature can be maintained at or above 72°.

5) Humidity can be controlled. Butyrate dope cannot be applied in high humidity, especially sprayed.

Fortunately, these features had been all provided for in the home shop of Mr. Ben Hall where the photos were taken of the Fleet 16-B re-cover work. In addition, a portable dehumidifier made it possible to work right through a rainy weekend. Wood frames support Pliofilm panels to make the enclosure in this basement; they extend from floor to ceiling. Ben also made the sturdy "spit" to support the forward end of the fuselage and this was of immense help. The wings for this plane did not need re-cover at this time. The ailerons were done because the wood structure required attention.

Having lived coast-to-coast, we can verify that some geographical areas are better than others for meeting the fore-going accommodations. For the most part, everyone can handle the covering and up through the hand-brushed clear dope at home (there is no particular fire hazard with hand-brushing dope), but a suitably lighted and ventilated shop is required for all of the spray work. With butyrate, temperature and humidity are critical.

Unless you expect to have frequent need of a compressor and spray-painting equipment, you can rent rather than investigating in purchase; they are not cheap and a poor one is most aggravating. The compressor must have sufficient continuous output to maintain 50-60 lbs. pressure at the nozzle, and include in the line a pressure regulator with a water-drain. If you rent, schedule the work so all the spraying can be done continuously (weather permitting).

For an envelope-type cover, another necessary item is a sewing machine, but this is usually no problem. A #16 size needle will handle the machine sewing thread, which is quite a bit heavier and more wiry than ordinary household sewing thread. It is *special,* not just heavier.

Other Essential Items

☐ One pair of sturdy, clean, saw-horses—high enough to be worked under, but not too high to reach over

☐ 1 6′ or 8′ measuring scale ("yo-yo")

☐ 1 pair of pinking shears (large)

☐ 1 pair of standard scissors (pinking shears for fabric only—regular scissors for all other things);

☐ 1 long rib-stitch needle (3-sided point)

☐ 1 large darning needle and thimble (for hand-sewing)

☐ 1 packet of straight or "T" pins

☐ 3 or 4 dozen wood, spring-type clothes pins (for clamps)

☐ 1 rubber suction-cup (plumber's friend)

☐ 1 household thermometer

☐ 2 good quality paint brushes: 3″ or 4″ and 1″ or 1½″, the smaller brush for applying lacquer-cement, etc., larger for dope.

☐ 1 gallon can with a top quarter cut out and the handle left on makes a good container for dope while working, or a 2 lb. can may be used though it is more tiring to hold.

☐ 2 five-gallon cans, wide-mouth opening with tight fitting lids, for mixing and strong Dope/fungicide and dope/silver.

☐ 1 or 2 dozen sheets of both #180 and #400 grit emery paper

☐ 5 rolls of ¾″ masking tape will take care of the entire job unless a complicated finish-trim requires much repeated masking. ¼″ and ⅛″ masking tape is available for pin-striping, etc.

Lay in a good supply of clean newspapers, and if floors must be covered, a couple of 8′ × 10′ canvases will be needed to protect them from dribbled dope, which is inevitable.

The same equipment will be needed for any covering materials used.

It pays to use the best quality of materials and to take the time to do a first-class job. It is too much work to do otherwise. It will last longer and it will be easier to take care of; it will certainly make a better impression wherever it goes. An FBO is largely judged by the equipment he puts on the line, and of course it will be seen at many airports other than his own. Neat, clean, well-maintained airplanes, regardless of age, speak well for his operations. A messy shop and shaggy looking airplanes scare customers away, and rightly so. In any part of aircraft work, it may cost more and take longer to do the job right, but it does pay off in the long haul in every way.

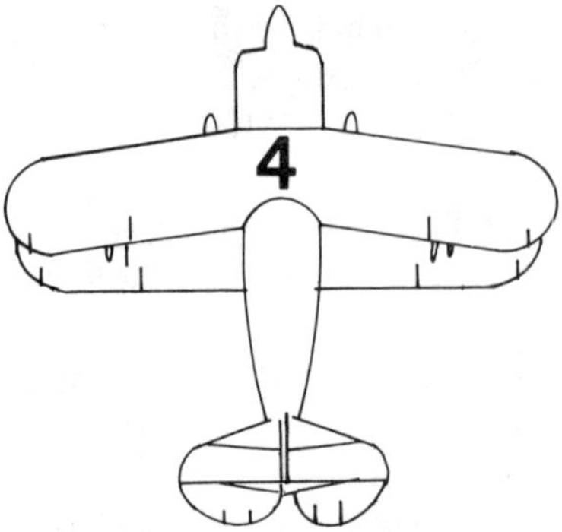

Preparations For Covering: Metal Structures

One of the most serious criticisms of so many re-cover jobs is not with the final cover and finish but with the lack of preparation *before* covering. There is only one reason for it (no excuse) and that is simply lack of integrity. How could any licensed mechanic sign off as "ready for cover" structures that haven't even had the dirt blown off, to say nothing of proper inspection and repairs? It is done; probably at the urging of an irresponsible owner who wants to get by the cheapest sort of way. The mechanic's "ticket" is on the line; he probably figures that if he doesn't sign it off somebody else will and he might as well get the few bucks. They are both gambling that there will be no accident with its subsequent investigation so no one will know the difference. Let's hope there will not be, but at best, someone down the line is being cheated.

The following is for the individual willing and able to do the work and get it ready for a good inspection prior to covering. Any structural repairs are to be done by a licensed mechanic *or* under his direct supervision. All work done by an unlicensed owner should be watched and checked frequently by his mechanic to be certain nothing is overlooked nor done in the wrong way. When in doubt, check *before* doing something, not afterward.

There has been a great variety of kinds and combinations of metal aircraft structure using fabric covering. The most common, of course, is the steel tube frame with formers and stringers (also of metal or of wood), usually having metal frame tail-feathers. The entire wing structure may be of metal—welded steel tube, such as the Stinsons of the 1930's; both spars and ribs were built up of tubular trusses like a bridge. Many of the light planes were a combination of steel tube fuselage and empennage with wings having aluminum ribs on either wood or aluminum spars.

There are still many planes in use that are all metal except for wood spars with the fuselage aluminum of semi or full monocoque type and the wings and tail surfaces fabric covered; for example, the Ryan ST-A and PT-22 series and the Meyers OTW's. Then there were the numerous all-metal-but-wings: the early Cessna 120s, 140s and 170s and the early Luscombes.

Bellanca still produces high performance 4-place aircraft which are all-steel tube frame, fabric covered, with their famous all-wood wing (also fabric covered).

The Navy-built N3N biplane trainer (originally, most were on floats) was built like a bridge, the fuselage structure being of riveted angle truss members (aluminum) with wings and tail surfaces also built up of aluminum. The planes were fabric covered except for the rear side of the fuselage which had removable aluminum panels held by Dzus fasteners, primarily because they were used in salt water and it was necessary to have access to the interior of the fuselage for cleaning and inspection. This was a convenient feature of these aircraft when used in agricultural operations, and some Stearmans have been modified to utilize removable side panels for the same reason.

Many of the large commercial airliners are still flying with fabric covered control surfaces. All high speed jets and turbo-props are entirely metal, as are most of the new light craft, but there will be fabric covered planes flying for many years to come.

Metalizing

Many people have been disappointed with the results of "metalizing" fabric covered airplanes. Their performance invariably suffers. The planes were designed for fabric which is lighter weight and sometimes even cleaner aerodynamically. The internal structure was not built for metal skin support nor was the airfoil designed for it. To achieve the same or better performance, a wing must be designed for metal skin. A good example of this is the Cessna 140. The original fabric wing gave much better performance than those which were metalized. The newer (later model) 140 which was produced with a redesigned, all-metal wing was a much improved airplane. "Metalizing" is a poor compromise. Weight goes up, therefore wing loading and power loading go up, and performance goes down. Increasing the horsepower to compensate just starts a vicious circle. If you want an all-metal airplane, buy one; don't try to modify a fabric plane. It will be spoiled and you will never be happy with it.

The first thing to be done in recovering an airframe is to *clean* all parts of the structure thoroughly after removing all of the old fabric. Remove all old paint and dope, oil, grease, hydraulic fluid, etc. Start with an air pressure chuck, if available, to blow off surface dirt; next best, wipe if off well with a rag. Wire-brush (hand) particularly dirty places, especially around the landing gear and its attachments and around the tail-wheel assembly.

It takes alcohol (rubbing alcohol will do) to dissolve hydraulic fluid. An old, small paint brush makes a good cleaning brush to get into tight corners, weld-beads, around bolt heads, nuts, etc.

Gasoline or cleaning solvent will remove oily dirt and grease, using a brush. *Warning:* work *only* outdoors with gasoline due to fire hazard. Solvent is much safer; it is also more expensive, but use solvent if you must work indoors. Both leave an oily film which can be removed with paint thinner or butyrate thinner; these may also remove the primer. However, this would probably be sanded or sand-blasted off later anyway, so that is of little concern at this point.

Rust

There is nearly always more or less rust on any steel tube structure and it can only be removed by hand sanding or, better yet, by sandblasting. The latter will thoroughly clean up places that are impossible to get into by hand, and especially weld-beads. It is very important to have clean metal in the areas of all welds for very close inspection for adjacent cracks which can develop along a weld or run out from it. For small parts of irregular shape, the use of a sandblast or pulverized-glass blaster is excellent for removal of rust. Prior removal of dirt and paint facilitates this operation and prevents the waste of much blasting sand or glass. Flat fittings and lengths of tubing without welds can be hand-sanded, but to be effective, all steel must be sanded to bright metal. In other words: *clean and clear of rust, completely.* Otherwise, rust will promptly reform under the primer.

A fuselage frame with extensive surface rust is best done outdoors with commercial equipment to get a first class job. Early in 1963 it cost us $60.00 to have a J-3 Cub airframe (including gear legs and tail frames) sandblasted, but it was well worth it. The equipment was rented to do the airframe (all except wings) for the

Fleet 16-B illustrated in following chapters. Most of the sand can be retrieved if the work is done in an enclosure, and used again another time.

Sand blasted parts must be sprayed with zinc chromate primer just as soon as possible to prevent immediate formation of rust (Fig. 4-1). Even though the primer looks good when an airplane is uncovered, sand off the chromate (small areas) in places particularly subject to rust to be certain the metal underneath is still clean and free of rust or corrosion (if aluminum).

Aluminum is never sand-blasted. Surface corrosion may be hand-sanded only with emery paper not coarser than #180 grit. If there is serious corrosion and pitting, that portion will most likely have to be replaced; this is definitely to be determined by a competent mechanic. Aluminum must be treated with a mild acid bath to remove all traces of oil, etc., before applying the zinc chromate primer so the primer will hold to the metal. While cleaning and repairing all parts take special care around rivets, bolts, seams, etc., for these are spots which hold dirt, oil and moisture.

Inspection

Before an airframe is re-covered, regardless of type of construction, inspection must be made for:

- ☐ Rust and/or corrosion
- ☐ Breaks, cracks or bends
- ☐ Loose bolts, screws, rivets, etc.
- ☐ Elongated holes at attachments of wings, gear, tail surfaces, engine mount, struts, etc.
- ☐ Worn bushings and bearings and control cable fairleads
- ☐ Bracing (rusted, corroded, bent or cracked)
- ☐ Control cables and pulleys (frayed, worn or rusted)
- ☐ Proper safetying of turn-barrel ends, studs, etc.
- ☐ Cotter keys wherever required (drilled bolts and castle-nuts are used at any rotatable connection except with ball-bearings)
- ☐ Proper length and type of bolt at every point

Replace any questionable wiring with correct size and type and run it through protective sheathing wherever it is to be enclosed and wherever it could chafe against any other part. Use rib-stitch cord to stand it off and to secure it at close intervals, and use "Adel" type clamps to secure it to steel or aluminum members. Leave no dangling loops of wire.

Lubricate all movable joints and levers; a light-bodied white grease such as Lubriplate is best for this. Oil will in time run off and dry out. Be sure that all controls operate smoothly and without excessive play. Wipe off excess oil and grease which would only hold dirt. Check again to make certain that *all movable controls* are fitted with castle-nuts and cotter keys, and that they are free to move but not sloppy loose, which would allow side play in the leverage.

The wing leading and trailing edge metal (usually aluminum) of older planes is most often both corroded and bent. It generally can be saved by some careful hand work, bumping out bends and dents and riveting in replacement sections that are necessary. Overlaps of the leading or trailing edge metal make less of a "hump" if made between ribs where P-K screws can be used to join them. File all edges straight and clean and round all corners. These pieces should be individually sprayed with zinc chromate before assembly.

Aluminum rib repair must be done by or under the supervision of a mechanic. They must all be sound and conform to the proper original contour. Any wing having flattened contour between rib trusses will not provide the intended flight characteristics. Reinforcing tape, tied securely at the root end of the wing is carried from rib to rib with a single wrap around each, terminating back at the root after a single wrap around the bow and returning to the root on the opposite surface of the wing. This reinforcement should be approximately midway between the spars. The wrap goes around each rib uniformly so that the tape runs on the *inner* surface of the cap-strips, so it will not be in contact with the cover fabric. We prefer the old method of carrying the tapes diagonally from the *top* cap-strip of one rib to the *bottom* cap-strip of the adjacent rib, etc. With a firm wrap at each cap-strip there can be no shifting in any direction and there can be no contact between the tapes and the cover. There is need of the inter-rib tapes only on wings of wide chord. The rib-stitching is going to prevent side-wise movement of the ribs, and where it does not it is time to overhaul the wing structure to correct the *cause* of the warpage or shifting.

Relative to this discussion of these inter-rib reinforcing tapes it might be well to state unequivocally that *all* ribs, regardless of the type of wing construction are secured to both the front and rear wing spars. There seems to be a common misconception that they are not. The wing is a sound unit in its entirety *before* fabric or metal skin goes on. It would be awfully flimsy otherwise.

After the components of the airframe have been inspected and signed off for approval to cover, put masking tape over all corners, edges, protrusions, etc., on which the fabric will rest. The better the protection, the longer the fabric will last. In Fig. 4-2, note the ¼″ masking tape used on the edges of the spruce stringers. Controls are installed and cables are tied out of the way. Clamps for the coaming, etc., are located but will later be rotated to fit holes without disturbing the fabric.

Drainage

Be certain now that drainage is provided for at all points. Drainage holes should be made in the bend of the wing trailing edge metal midway between each rib of all wing panels, because moisture can collect in that V, with subsequent warping and rotting of the trailing edge tips of the ribs, although this is much less likely to happen with seaplane type grommets. These holes should be ⅛″ dia. and should be filed across the apex with a small round file. Do not use a corner of a flat or triangular file as that would result in a sharp corner top and bottom, where cracks could develop. Every hole in metal must have radiused corners and small holes must be *round*. Make these holes toward the lower surface, that is, in the

Fig. 4-1. Sandblasted fuselage frame and parts (hanging).

Fig. 4-2. Controls installed, wires and cables tied out of the way; stringers installed with ¼″ masking tape covering edges; fuselage frame ready for cover.

lower radius of the bend. Clean away any burrs around the edges of the holes so they are clean and smooth inside and out. These are advisable on any wood wing.

One of the very few AD's on the PT-series Stearmans calls for holes in the lower surface of the metal at the bottom of the aileron wells where moisture collects and rots ribs and the bottom edges of the rear spars. No. 10 holes are to be drilled between ribs. These, as all others, are to be cut cleanly open (*not* punched) following all doping.

5

Preparation for Covering Wood Structures

We approach the subject of wood aircraft structures rather hesitantly, not from personal lack of experience (the writer had an Approved Repair Station Rating for Aircraft Woodwork for a number of years), but because of the common attitudes toward it. Very few licensed mechanics have had any experience with it and will usually tell you so, whereas the aircraft owner, also without *aircraft* woodworking experience, will all too often tackle the repair or rebuilding of an airplane wing or other component on the assumption that he does know all about it—"it's just wood, isn't it?" There is a lot more to it than that. Some pretty bad errors result. Cabinet workers are excellent woodworkers but they do not know anything about building airplanes. I've rebuilt a lot of wings, but do a lousy job on cabinets, etc. Aircraft work is *all* specialized.

First again is the question of legality, which greatly concerns the FAA. Then there is competence, no matter *who* does the work. And here we are in the same situation as with the dope and fabric work, but this is more serious because we are dealing with the basic airframe structure.

How It Shall Be

Back to the book. Chapter 1 of *AC 43.13-1* spells out *how it shall be*. What it does not tell you is how to do it. This manual is for mechanics who are supposed to know how. No reflection on today's mechanics; they are not taught and have had little opportunity to pick up this experience. As you know, *only* licensed airframe mechanics or other appropriately rated personnel are permitted to make major repairs on ATC'd aircraft, regardless of type of construction, and wood is no exception. It is just as easy to get a bad glue joint as a poor weld joint and it is no less serious. We advise consulting your FAA office about your particular airplane and just exactly what *you* will be permitted to do on it, all circumstances considered. It is here that your own integrity and history of relationship with your FAA men will be a big factor. Under any circumstances, you will be under the supervision of your licensed airframe mechanic. Regarding any repairs, it always pays to *ask first*; it can save you a lot of tears later.

For the limited purpose of pre-cover preparation, only minor repairs will be discussed here. Major wood repair and rebuilding would require a book on that subject alone.

The owner of an amateur-built aircraft does not have the limitations imposed upon the owners of production airplanes. We will not become involved with the possible injustice here; there are valid arguments both ways on the subject. We will assume, in this instance, that the amateur-builder/owner knows what he is doing on his own plane. Since the topic is being limited arbitrarily to preparations to cover or to re-cover, it applies the same to all.

Wooden Wings

Let us begin with the most common and simplest involvement with wood in a typical light plane having an all-metal airframe except for wood wing spars. Unless these spars have been replaced within recent years, they are probably over thirty years old by now. This does *not* mean they are no longer airworthy. It does mean that they should be thoroughly inspected by someone who knows what to look for—especially *under fittings*. This requires the removal of all metal. Particular attention must be given to wood under fittings because all too often no varnish was put on these areas before assembly of the wing, so there was no protection from moisture where it was most needed, although this is most often found in the case of a replacement spar than with originals. But varnished or not, inspect closely for cracks, checks and compression damage, especially at bolt holes. Check both sides of spars. *None* of these defects are permitted. The Fleet aileron fitting area shown in Fig. 5-1 is an example, where the boxed portion supporting the control arm fitting and one of the three hinges (on both ailerons) had not been varnished inside at all. Fortunately, the wood was still sound, thanks to snug-fitting fabric. However, all of the hardware was badly rusted. It was necessary to open the boxed

Fig. 5-1. Repaired Fleet aileron with heavy ties holding upper part of the aluminum leading edge metal in place for nailing.

section to refinish the inner parts and to replace one top cover piece (spruce). The aluminum hinges were corroded and their steel bushings were worn. In fact, about half of them were just gone.

Any partially loose plywood (or spruce) spar reinforcing plates should have glue worked into the gap with something very thin (such as a razor blade), then clamped tightly closed using wood blocks and large clamps. Do *not* attempt to peel off loose plates; more damage is likely to be done than not, and you would be getting involved in major repairs. *Do not add nails to secure reinforcing plates.*

If the spar is in good condition, sand it and then re-varnish it. Two coats of marine spar varnish of best quality should be used on bare wood; one coat will suffice on previously varnished wood. If it was necessary to completely disassemble the wing to work on the spars, this is certainly major repair. Assembly will be discussed farther on, along with the all-wood wing.

Any rust or corrosion on any wing fittings must be removed as described in Chapter 4. Any bent fittings must be straightened (if feasible); all are to be closely inspected for cracks, particularly around welds. All, whether steel or aluminum, are to be zinc chromated, whether or not they originally were. This applies also to the leading and trailing edge metal, especially on the inside. A good dope and fabric job should be expected to last for long years so all parts must be properly protected.

A wing is assembled upside-down with ribs loosely in place. The compression members and their associated parts are fitted first. *Never over-tighten any bolts in wood.* This is one of the most common and most damaging of errors committed with wood aircraft structures. Bolts through wood are *never* torqued as in metal because wood is compressable. Some spars have certain bolt holes bushed, but even these are not to be tightened as would be done with metal. Over-tightening crushes the wood, weakening it considerably in that area. Tighten nuts enough to be firm; the bolt should not rotate readily; that is enough. This applies to all bolts throughout the wing without any exception, and to all wood aircraft structures.

The internal drag and anti-drag wires go in next. There are a number of kinds of end-fittings; some very old planes had turnbuckles for adjusting tightness; nearly all have some version of screw-end. Those having left and right-hand threaded ends for fork-end fittings should be put in so that all right-hand threads are at front and left at the rear, or vice-versa, but uniformly. It is very annoying and time consuming to have them mixed. I always have made a habit of putting all right-hand threaded ends at the front spar.

Mark precise center-points at the centerline of each spar where it intersects with the projected centerline of both fore and aft ends of each compression member. The *exact* spacing of distance between spar centerlines *must* be maintained. Due to possible slight shrinkage of the spars, it may be necessary to shim at one end of one or more compression members to get the specified spacing; this should not exceed 1/32″. The wing cannot be accurately trammeled (which is simply "squaring") unless this measurement is constant the full length of the wing.

If someone is available to help, it is possible to trammel a wing by using a steel measuring tape, but this permits too much error. A set of trammel points on a steel bar (about 6 ft. long) is much more accurate and easier to use.The diagonal center-line points are used. Starting at the root end of the wing, set one point at the root compression member center-line point of the front spar and adjust the second point to fit the mark at the rear spar of the next member. Then reverse and measure the crossing diagonal. Adjust the tightness of this pair of wires until this pair of diagonals is equal to each other. *Do not over-tighten* these drag and anti-drag wires; they should be quite firm, but no more. The wire in the "short" direction can be eased out (loosened) to equalize the measurement just as effectively as to tighten its mate. Proceed in the same way for each bay or pair of wires. Depending on the plan-form of your wing, it is possible or even probable that these internal wing bays will not be the same dimensions. The compression members are usually not the same distance apart from each other; therefore, the lengths of the pair of the internal wires will differ from one bay to another. All that is necessary is to have each individual bay squarely adjusted and *all* of the wires as nearly the same tension as possible.

Over-tightening the wires not only pre-loads the wires excessively but can also pull the wing out of shape and twist it to the extent that it could never be properly rigged on the airplane. A completely assembled wing should rest flat. By lifting one end only it can be easily twisted by alternately raising or lowering the front or rear spar end, but this is normal flexibility. When it is mounted on the fuselage, with all struts and wires attached, it will be quite solid. This flexibility is one of the greatest advantages of wood construction. (We make note here that a fully cantilever, stressed-skin wing is not this flexible.) Be certain, before securing locknuts or other locking fasteners on the internal wires, that the wires themselves are not twisted. It helps to lightly oil the threaded ends as they are installed so they will turn freely without binding, then hold the wire with a suitable tool (that will not scar the wire) while tightening the locking device.

Use nails no larger than #20 × ½″ for securing the aluminum leading edge metal to the top and bottom edges of the spars (and to the ribs of an all-wood wing). P-K (sheet-metal) screws are used to attach it to aluminum ribs. If formed of two sections (top and bottom), it is first attached to the leading edge piece, working the metal back and aft over the ribs. If it is formed of one continuous piece, nail it first to the bottom edge of the front spar, then turn the wing over and nail (or screw) it back to the top edge of the spar. A "half-hard" aluminum is generally used for leading edge material and it is rather tough and springy. It must lay snug in contact with the rib leading edges to properly conform to the airfoil, so it is usually necessary to use some form of persuasion to hold it tightly to the ribs while attaching it. See Fig. 5-1 showing how straps of tough fabric were used to hold the leading edge aluminum to the Fleet ailerons for nailing down the upper surface.

Speaking of an "all-wood" wing, we mean one having the basic parts—both the spars and the ribs—of wood. Compression members, tip bow, and other internal parts may or may not be of metal. Since this is intended as a guide for covering, only the most common minor repairs will be considered here. We will assume a commonly used composite structure is in normally airworthy condition. Certain ailments are very often found in nearly all wings due for re-cover.

The spars were discussed in the preceeding paragraphs so let's look now at the wood ribs, very often warped and with loose gussets. Refer again to *AC 43.13-1, Figures 1.10* and *1.11* for rib repair. If the wing generally is in bad shape it should be completely rebuilt; if all ribs are bad, it is actually easier to build a full new set than to do extensive patching, and it is much more satisfactory. But if only loose gussets need to be replaced, that is not difficult.

For any repair, use only *aircraft-grade* materials; spruce, plywood, glue and nails. Other are *not* "just as good." Replace gussets *as per original* in dimensions, grain direction and type of wood (birch or mahogany plywood).

Use aircraft nails: flat-headed, cement-coated and barbed, not exceeding #20 size in any rib material nor #18 for any other places. Steel nails are preferable to brass because the heads are flatter and thinner and they are stronger; brass nails are soft and bend too readily, especially when going into plywood. The length of the nails is important and so is their spacing. They need to be as long as practical for good grip, but *never* go *through* the base wood. For gussets on ribs, if nails are used that are the length of the width of the cap-strip or trusses, the thickness of the gusset will be the margin not penetrated, so this length would be safe. The *width* of the rib material is not always equal to the *thickness*. A common error is to use too many nails per gusset; see *AC 43.13-1, Fig. 10*. Always keep in mind that an equal number of nails is being put in from the other side too. There will be a nail at each corner of the gusset and possibly several along the greatest dimension, besides others depending on the number of truss pieces at the cluster. Use no more than necessary, but enough to bond the gusset uniformly.

To handle the very small aircraft nails, I use the magnetized end of a light tack-hammer to initially place the nail, then turn the hammer and drive (set) the nail firmly using the main head. A few light but solid strikes sets the nail much more securely than numerous little tappings. To drive nails into ribs which are assembled on a wing, the rib must be solidly backed with a steel bar, such as a bucking bar or block about an inch thick and of a size easily hand-held, held in the other hand firmly against the rib; there must be the least possible deflection of the rib while working on it, to prevent further damage. To work in close, tight corners, such as in next to a spar where the nails cannot be struck solidly with the hammer head, use a ¼″ diameter punch struck by the hammer to drive the nails. Do not strike nails so hard that the wood is crushed; there should be no imprints of the hammer. To hold a pair of gussets in place temporarily until enough nails are in place, use a small C-clamp, then remove it and continue nailing. Aircraft nails are barbed so they hold tight. *Do not pull nails* in the mistaken idea that this is in any way beneficial. Properly set, nails *do* hold permanently; many the old wing I have repaired where *only* the nails were holding things together. (The old casein glue was long gone to dust.) Also, far more damage can be done by pulling out the nails than any possible advantage in leaving them to do their job. It is not likely that there could be even one pound of aircraft nails in all-wood plane of normal size. A wing with plywood skin is something else again; the skin is put on by the use of external nailing strips to provide even pressure while the glue sets, then these are removed along with the nails which hold them. That is the only time nails come out purposely. For the necessary removal of nails to do repair work, I use a pair of very small, fine diagonals, purchased from Snap-On-Tools.

In any work with glue, put it *just* where it is to go, and after the joint is secured by nailing or pressing, wipe off the excess as soon as possible, before it has time to soak into and dry on surrounding wood surfaces. No glue should be left exposed on wood anyplace. It is not only obviously messy workmanship, but eventually it would flake away and leave bare wood exposed to moisture and dirt.

If any ribs are loose from the spars, work glue into the separations with a very thin probe or razor blade as thoroughly as possible and re-nail. Again, use no nails larger than #20 and of a length to penetrate no more than half the spar thickness for the vertical members and not exceeding ¾" total length to secure the top and bottom cap-strips to the top and bottom spar edges. You have probably been told or have read that rib cap-strips "should not be nailed to the spars." We have yet to see a wood wing constructed (ATC'd types) where ribs were not glued and nailed to both spar faces and edges. Trouble comes only when too large and/or too many nails are used. Relative to this, ribs are *not* to be varnished before assembly to the spars; glue will hold only on clean, bare wood. The only varnish used before final assembly is completed is to be put on the spars at the areas of fittings, both sides, a minimum of two coats.

Removing Old Glue

When refinishing old wings which have old glue showing, which was not cleaned off when they were built, sand it off clean. Most of them do have this residue—or bare old wood where it used to be. These places, the bottom edges of the rear spars where the ribs attach and the trailing edge tips of the ribs, are the most common places for dry rot to develop with resulting deterioration in the strength of the wood. Laminated wood wingtip bows suffer too, with the laminations separating and the bow warping inward between supports. All of this is the direct result of unprotected wood combined with lack of adequate drainage and ventilation. All wood must be clean and dry. It is tedious work, but sand all of the wood, and especially clean up the hard-to-get-at places, because they are the most likely to collect dirt and moisture. Sand edges, but just enough to round slightly; do not remove appreciable amounts of wood. Remove burrs from edges of gussets. All wood for rib-truss members should be machine-sanded to provide smooth gluing surfaces. Do not hand-sand faces of a glue joint because there is too much likelihood of unevenness resulting in a poor mating surface. A disk or belt-sander will produce a clean, straight surface. Sawdust from any sanding *must* be removed before gluing and also before varnishing. Use of air pressure, if available, is the best way to get it out of particularly enclosed "boxes" and the numberous pockets formed at all of the many gussets, etc. A medium or fine grit garnet paper works best; I like Behr-Manning 100C-2/0 garnet paper for hand-sanding of wood, both spruce and plywood. Use a wood block to back the paper when sanding flat surfaces. Garnet paper is far superior to ordinary sandpaper; it cuts better and lasts longer.

The accompanying photo (Fig. 5-2) of the pair of Fleet ailerons illustrates several points in addition to the unvarnished fitting areas previously mentioned. The one toward the back of the work table is right-side up; it was in better condition than the one in the foreground. The latter is shown upside-down. The wood trailing edge was removed; the aft tips of the ribs were cleaned up and their wedge-blocks replaced. The end rib (extreme lower-left) had been replaced sometime in the past, but not varnished; the Resorcinol-type glue still is all too obvious. A new spruce rib was put on the other end (out of view at right). C-clamps are holding the aileron to the edge of the table to line it up for placement of the trailing edge (Fig. 5-3). It can be clearly seen why varnish must be hand-brushed into the many deep pockets, especially at the trailing edges of such parts.

Wood aircraft structures must be varnished by hand, one side and then the other, in order to get the varnish completely into all the deep angles, pockets and corners. The inside of all closed box sections must be varnished before the final closing piece is put in place, which means that this piece must have its inner surface varnished. Wherever it is to be glued and nailed has to be predetermined and left clean so the glue will adhere and make a secure joint. As varnishing proceeds, keep going back to brush out varnish from the low spots to which it runs, so there will not be thick, sticky droplets. You want uniform coverage; do not brush it on thick and heavy, but thin and evenly. It will not dry where it is too thick and there would be bumps and ridges under the fabric.

Older wings which still have a fairly good varnish cover will need only one coat following sanding to clean them up. But on any raw wood, new or old, a minimum of two coats are essential. The first will soak in completely. Never use cold varnish; it should be at least room temperature. Allow it to dry before adding the second

Fig. 5-2. Repairing the wood ailerons; warped trailing edge and gussets removed; boxed portions opened for sanding and varnishing. There are spruce, wedge-shape filters at trailing edges of ribs.

coat, which should give a gloss finish. If it does not, use a third coat. A third coat is always advisable on and *in* plywood-boxed portions.

Use only best-quality exterior varnish; we use a good marine spar varnish. It penetrates well and is highly protective; it is durable and resists "lifting" under doped fabric. If a little added weight is of no concern, white dope-proof paint areas to be in contact with the fabric, but this is not always infallible.

Wing-mounted airspeed lines, light wires, etc., will be reinstalled after varnishing. Rib-stitch cord is used to stand off and secure wiring and this should be shellacked to prevent deterioration. Be sure that all such items are clear of any control mechanisms and that the latter have been inspected for good working condition and smooth operation. This, of course, is the duty of the A&P, but you can save him a lot of time and trouble by doing the work—as always, under supervision. (We keep emphasizing this to keep you out of trouble.)

While the wing is still open, make a sketch with accurate dimensions for the proper locations of all necessary openings for inspection-cover rings, access holes and drain grommets. Do this for all components being covered. Once the fabric is in place, it is difficult if not impossible to determine all these places without overlooking some of them.

Look closely at the photos of the completed Fleet ailerons, Figs. 5-3 through 5-5. In the first in foreground, the new end rib is obvious, as is the new bottom "box" piece by the large clamp next to the far end. The small C-clamp holds the trailing edge strip (spruce) until glue dries.

The next picture (Fig. 5-4) shows the wood work completed and varnished and reconditioned hinges re-installed; the actuating lever is out of sight in both, but shows in the following picture. Figure 5-5 shows the completed ailerons ready to be covered. The masking tape covers the nail-lines over the ribs' leading edges where the aluminum is nailed to them, and around the edges of the metal. Photos do not show them, but ¼″ holes were drilled in all bottom, corner gussets at the trailing edges. Had this been done previously, *none of this repair would have been necessary*.

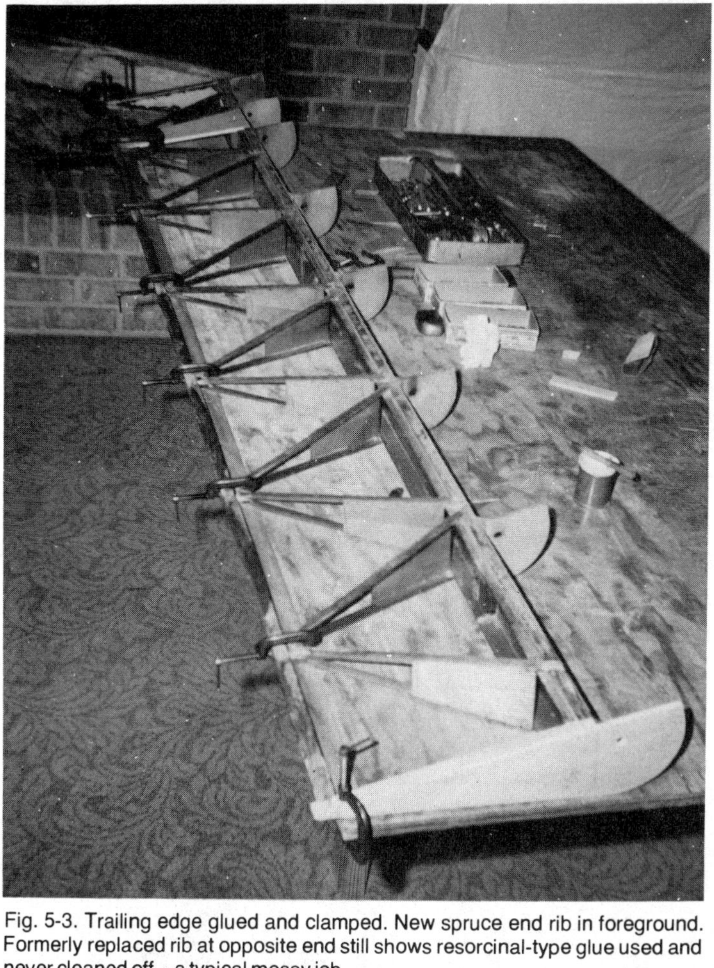

Fig. 5-3. Trailing edge glued and clamped. New spruce end rib in foreground. Formerly replaced rib at opposite end still shows resorcinal-type glue used and never cleaned off—a typical messy job.

Fig. 5-4. Aileron repair completed. Reconditioned hinges replaced. Ailerons varnished (two coats) and ready for the leading edge aluminum to be replaced. (See Fig. 5-1.)

If you are preparing to cover a newly built airplane, by now you have begun to appreciate some potential problems and can realize the importance of many details. The best way to plan a new airplane is to study the maintenance and operational problems of old ones of all kinds.

A well constructed and finished wood structure built of approved aircraft grade materials will last indefinitely with proper care and maintenance, and *maintenance* is the essential thing for any kind of structure. Wood absorbs vibration and shock to a much greater degree than metal; it has more elasticity and absorbs noise better.

Wood Glues

There are excellent new wood adhesives (glues) on the market today, but as compared with the Weldwood plastic resin (water-mixed) glue, which we use exclusively, we feel they have disadvantages that outweigh their possible advantages, not considering cost: more critical in mixing proportions, working time and atmospheric conditions, difficult to remove messy excess and difficult to remove for repairs. The Weldwood has excellent qualities of adhesion, it is highly water-resistant and durable; it is not overly critical to mix and its "working life" extends up to four hours. We have never had a glue joint fail with it. Claims for certain adhesives intended for aircraft use that will bond without pressure are highly questionable indeed. The object is to *bond* the two pieces of wood, not just stick them together. A proper glue joint *must* be cleanly cut to achieve evenly mated and bonded surfaces; it requires an even and constant pressure during the setting time of the glue at a temperature favorable for drying (minimum 70°F). The glue line of a good bond shows only as a fine hair-line—no more. Anything else at a glued wood joint is the equivalent of soldering compared to welding 4130 steel—it is *not acceptable*. You don't do it out in the garage in winter time. You do need proper tools.

Only gussets are secured (pressed) by nails alone. All reinforcing plates on spars and all scarf-type joints are to be backed with wood blocks of sufficient size to cover the parts (area) being glued together, and clamped with C-clamps or parallel clamps for 12 to 24 hours at room temperature. Never use clamps without wood backing blocks to evenly distribute the pressure and to prevent the wood being crushed by the clamps. Use thick enough blocks that they will not bow outward between clamps. Do not crush the wood.

A small brush works well for applying glue to large areas, such as extensive laminations or putting on plywood skin, but for regular use where only a small amount of glue is to be put on for gussets,

Fig. 5-5. Fleet ailerons ready for cover; masking tape over all edges of metal, which was zinc chromated on the inside.

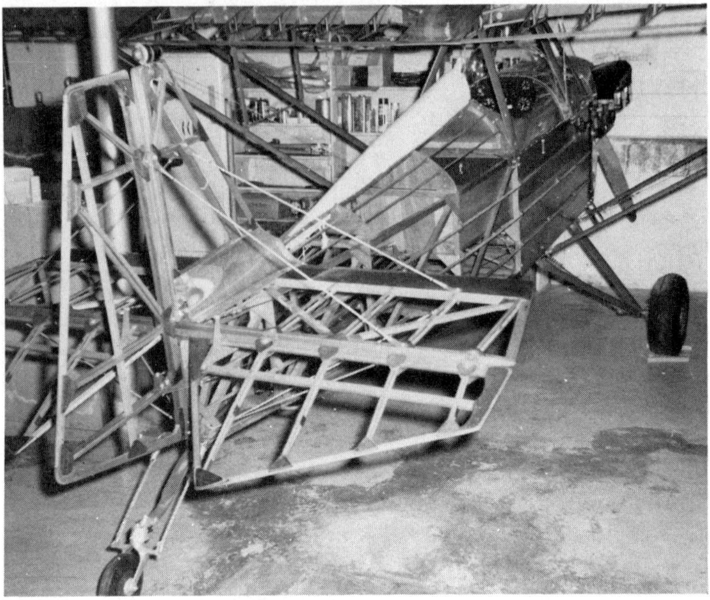

Fig. 5-6. Completely assembled and rigged all-wood homebuilt ready to be covered (after disassembly). Note fabric on headrest.

etc., a stick sanded to a flat taper works very well and it can be repeatedly re-sand to remove excess dried glue.

For both wood work with glue and in working with dope, use expendable paper towels instead of rags.

The grain of the wood in all cap-strips is to run *vertically*, not *flat*, to naturally resist warping flat between the rib truss with consequent loss of rib contour and the wing's proper airfoil section which largely determines the flight characteristics of the airplane. Lightly sand all parts of the wing, or any wood component to remove burred edges and surfaces which hold moisture and dirt and result in eventual deterioration. A clean, neat job is desirable from every standpoint. A messy job not only looks bad. It is bad.

The series of photos of the repair and the re-covering of the Fleet 16-B ailerons illustrate common problems involved with old aircraft surfaces as well as the covering procedures described in Chapter 8. After removal of the fabric, the condition found was typical. There had been previous repair, not too well done. The one prime cause of trouble was the usual lack of adequate drainage—no drain holes in the trailing edge, bottom-side gussets, and poorly cut out drain grommet holes in the fabric. As mentioned before, no varnish had been put inside the boxed parts nor at the repairs done at some other time. This was in contrast with the fuselage which was in excellent condition following sandblasting. Perhaps the previous owner just didn't know aircraft woodwork? The wings did not require new fabric at this time, but they will be re-doped to match the rest of the plane. Incidentally, the Fleet biplane wings have wood spars and aluminum ribs; the upper wing is one-piece.

The all-wood home-built shown in Fig. 5-6 completely assembled for rigging prior to covering, was built by the writers; I did the easy part: the wood structure and everything but the welding, but you would never believe the amount of fittings, struts, etc., that go into one little "all-wood" airplane, and W. E. S. did nearly all of that—*several times*, because I have had a lot of bad luck with it. And I can say from experience: the wood takes stresses (and repairs) far better than the steel parts. Never underestimate wood, but take good care of it and it will take good care of you.

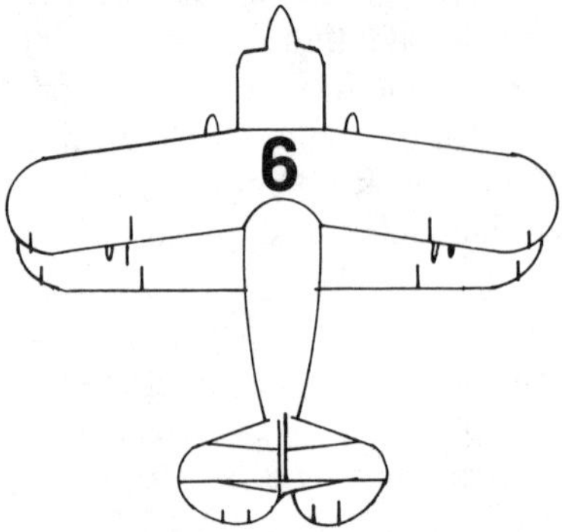

Interior of the Cabin or Cockpit

It is beyond the scope of this book and our own personal experience to become involved with cabin upholstery. This will deal with very elementary covering inside the cabin of light planes, using the same fabric with which the airplane is to be covered.

Why Cover Interiors?

There is an advantage in covering the interiors of open cockpits, not as important from the aesthetic standpoint as from a very practical one: it prevents weather and sunlight from deteriorating the *external* fabric *from the inside*, particularly if a cockpit cover is not used. Of course, if the inside is nicely finished with or without fabric a cover is an absolute necessity, but there are many times when the plane will be sitting out in the sun without the cover on. Some owners have avoided this problem by using aluminum instead of fabric from firewall aft to behind the cockpit. This also saves fabric damage from feet and knees—no small item to consider. But doing this would depend on how "original" you may wish to keep an old airplane, or perhaps weight on a small plane or the ease of entry and exit (the harder to get in and out of, the most likely the damage). If an airplane must be tied down outdoors, a good cockpit cover is a *must* in any event. The cabin airplane has a clear advantage under these circumstances. If the aft end of an open plane can be reasonably well closed off behind the cockpit it will be less drafty and more comfortable. Anything that keeps out moisture and dirt from the structural and working parts is all to the good.

Because the frame structure within the cabin areas of different kinds of light planes varies greatly, only a general outline will be given. The primary concern is neatness. It is desirable to leave exposed only the controls, while making the most snug closures around the inevitable openings essential to their operation.

Your airframe will have been inspected before covering begins. Therefore, it is assumed that all wires are properly supported and protected, also all control cables and tubing for all systems. Some of these items will have removable metal covering plates for access, but many things will be enclosed between inside and outside fabric which would require cutting into for repairs. So be certain everything is properly supported, protected and lubricated for long use.

The entire cabin area has been stripped to bare, basic airframe: seats, doors, window moldings, baggage holders and every other item. If the airplane is new, this is the place to start your covering job, and the interior fabric precedes the installation of all these things. Either old or new, this is the place to start fuselage cover. Those readers having open cockpits without interior fabric may be excused from reading this section, and we will go ahead for those wanting at least a plain, neat and attractive "office."

Starting overhead, use clothespin "clamps" to hold a piece of fabric, which you are certain will be a little larger than required, to the front crossmember(s). Using hand-sewing thread and needle, make temporary ties to hold it up to the overhead members as you work aft, then securing it from the center, outward to each side. Whether or not the top piece can be carried on down the rear part of the cabin continuously will depend on the structure and provisions for fuel or baggage behind the seats. When it is fitted straight and evenly, use the hand-sewing thread to secure it to the primary overhead members and wherever it must be supported in concave contours by taking the thread around the underlying parts with two or three laps of thread, then typing with a square knot and clip the ends. These knots will later be covered with neat pinked patches. It is not necessary to make many of these ties—ten or twelve inches apart across the top will be sufficient. Use them also on side sections where there may be extensive unattached areas of fabric.

Before starting to use the lacquer cement to attach the fabric, mask off everything that is left in the cabin, such as the instrument panel. Yes, it can be removed, but like the butyrate dope, it would probably take off paint or varnish with it. Also, protect the fabric from lubricating oil or grease on controls and cables.

Cut the front edge straight, leaving about one inch allowance to be cemented up and around the tube member. Apply the cement with a small brush and keep it on the surface of the tubing *away* from the side being covered. If this overhead fabric is *not* carried on down the back of the cabin, cement it likewise to the top rear member.

With headliner secured at top, front and rear, trim and clamp around the contours of the side windows and doors, again allowing about an inch for cementing area. Cut notches in these allowances for neat contours around bends and corners to avoid bunching of the material inside and over-laps on the outside. These places will be overlaid with the exterior fabric and with the finishing pinked tape, so unnecessary layers must be eliminated.

The cabin side fabric can usually be carried forward to the fire-wall in spite of obstacles. With a fuselage fuel tank and the many controls, this can be difficult at best, but it is desirable to have it all neatly enclosed. Be certain that everything is *clean*. It will be necessary to make a number of cuts and holes in the fabric for wires, cables and controls in most light planes. Keep these as neat, short and small as possible. After cementing the bottom and front edges, close the cuts with a few small stitches so there will be no extensive gaps when the fabric is stretched taut.

Carry the side pieces aft of the rear seat continuously if possible. If there are doors, it must be put on in sections. The fewer the separations, the neater and less drafty the cabin will be. Use the lacquer cement to attach all edges to longerons, lower window edges (as the upper portions), door frames, and front and rear vertical airframe members.

It is *not* necessary or desirable to pull the fabric tight. Concentrate on keeping it straight and even. Do not worry about a few wrinkles, but fit it so there will be no folds in curves and corners. The better all-around closure you have, the more comfortable and weather-proof the cabin will be. Cement all edges well for long lasting good appearance. Fabric coming loose at corners and edges always looks "doggy."

Doping The Interiors

The interior will be finished the same way the outside will be so the doping procedure will apply the same to all parts. When working with linen or cotton, it begins with water-shrinking; the synthetics are shrunk with a hot iron. A clean turkish cloth (wash cloth) works well to *thoroughly* wet the entire area of the fabric with the distilled (or mineral-free) water. (Most tap-water now contains too many minerals, etc.) Have the cloth *wet* and rub it in uniformly to completely wet the fabric fibers. Suddenly, as it dries, it will look just great as all the sags and wrinkles disappear. There will be small gaps where you had to cut out for controls and things, but these will be covered with pinked tape. Allow 24 hours for the water to dry out completely before starting to dope.

It is as important to use an anti-mildew protective agent on the cabin fabric as on the outer cover and the mixture will be enough for the entire airplane unless it is a very big one. The fungicide comes in one pound cans (under various trade-names). Since the first coat of dope must be brushed in very well, it will be thinned much more than subsequent clear dope coats, about 50% thinner to dope. To mix, put the anti-mildew paste into a wide-mouth five gallon can which has a good lid for tight closure between periods of use. Next, pour about a quart of butyrate thinner in and stir well until the colored fungicide is completely dissolved. This material is a blue or bluish green color and it is permanent. Add the dope and the balance of the thinner and stir thoroughly. As in painting, it is necessary to stir the mixture frequently to keep the more solid fungicide from settling to the bottom. Brush this mixture in thoroughly and uniformly, but *not* so much that it runs, inside (or in this case, the outside). One application of the fungicide is sufficient. Use the wide dope brush for full coverage.

The second brushed coat of clear dope, as those following, is thinned only enough that it can be spread evenly. Do *not* brush back and forth vigorously as you would normally do when painting; dope thickens and dries rapidly and excessive brushing results in a rough surface and uneven spread. It should go on as heavy as possible but without runs or puddles. The object is always an *even* and full spread. This inside cabin area is an excellent place to practice doing the outside which is more exposed to the critical scrutiny "of God and everybody." It is also the most difficult because of the somewhat limited working space. If you get a neat job on this you are off to a good start.

After this second coat the fabric usually looks and feels like an old paper bag. Don't despair. One more coat and it will be noticeably improved. There will not be much "fill" of the weave yet but it will have begun to tauten. When the third coat has dried, sand it completely. There is always a slight build-up of dope on the nap of the cloth. The more frequently it is sanded, the smoother the final finish. The #180 grit emery paper is best here.

With the fabric now tautening, it is time to put on the pinked tapes. The 2″ width is used. Start by covering any cuts you had to make. Round all edges that do not line up square with an edge and make neat round patches to finish holes for control rods or cables, wires, tubing, etc. One inch diameter circles will nicely cover the tie-knots which secure the fabric to the top and side frame members. Put tapes wherever there are underlying structural members, and then the finishing lengthwise tapes *except* those which will be overlapped by the outside covering fabric; this includes all edges of windows and doors which will have the pinked tapes over the outer covering and we want to avoid extra thickness. In other words, do not leave any raw fabric edges exposed, and reinforce any place the fabric *does* or *can* contact any underlying part.

When applying pinked tape, always watch to make certain that it lays straight or follows a contour evenly. Brush dope on quite heavily to lay the tape. It can be lifted and reset as necessary as long as the dope is wet. Apply the dope only as far ahead as you can keep up with laying the tape; this will vary, depending on the shape or surface being covered. Brush over the tape with a full load of dope as soon as it is in place. The dope dries as it is in place. Sometimes it is necessary to work it into hollows with the fingers until it completely contacts the base fabric underneath. Do not permit such gaps to dry because once the tape begins to tighten it will remain stretched over these hollows. Do *not* pull the tape over any concave sections, anywhere. It *can* be pulled firmly and to advantage over *convex* and corner areas, where the pulling will at the same time assist in straightening it and also help lay the edges. Once the pinked tape is on and given a coat of dope it is there to stay. Apply another coat to bring up the fill to match that of the base fabric, then sand it with particular attention given to the pinked edges before continuing additional over-all doping of the entire covered area inside the cabin or cockpit. Photos in the following chapter illustrate putting tapes on the outside, and of course the procedure is the same for all parts.

Six full brush coats of clear dope, thinned only enough for even spreading and with light sanding between every other coat, will produce a nice smooth base for the silver.

Spraying the silver dope does a much nicer job, but with the very limited access inside the cabin, it may not be feasible. Furthermore, at this stage you may not be working in a place suitable for spraying. So we will assume the interior will be hand-brushed, at least in the inaccessable places. Hand-brushing the silver *does* always leave brush marks and never looks as nice as a sprayed surface.

Mix the silver much as you did the fungicide in a wide-mouth gallon can. Silver paste is preferable to powder. A half pound of paste is best mixed with ½ gallon of butyrate thinner—very throughly—then add about 3 gallons of dope and stir well. This is a heavier mixture than would be used for spraying, which would require the addition of another gallon of thinner.

Four coats of silver should give a nice finish, again with light sanding every other coat. If it is to be left silver, do the final sanding with #400 grit emery, then a last coat of only partial-silver mixed with about 50% clear dope. (More regarding this in the following chapter.) Any color trim is best done at the same time it is put on the balance of the airplane because it is very difficult to get a perfect color match otherwise.

Interior Trim

Planning the interior trim will depend on the individual's taste. It is nice to use a matching or contrasting color to complement the exterior, but restraint is adviseable. It is better to avoid flashy distractions, especially in the vicinity of instruments and controls. Brightly colored or reflective trim has no place in the pilot's range of vision. Flat instead of gloss finishes are recommended for the interior open, structural parts, such as the tubular members in the cabins of J-3 Cubs and 7-AC Aeroncas, etc. It can be attractively finished without being distracting.

Should you wish to use insulating material around the cabin area between the inner and outer cover, get approval and advice from your mechanic if you are working on the ATC'd airplane. Run all controls, cables and wiring in suitable protective covering of permanent quality. Have this inspected and approved before enclosing it.

You may elect to proceed with the outside covering as soon as the clear doping has been completed in the cabin area and then do all of the silver doping on the fuselage at the same time. This is recommended. Just make certain that no dirt of any kind, especially oil, gets on the clear-doped interior.

Do not leave masking tape on a painted surface for any extended period of time; it could remove paint when pulled off. If you can continue working on the covering without interruption, leave the instrument panel, etc., masked until work is completed.

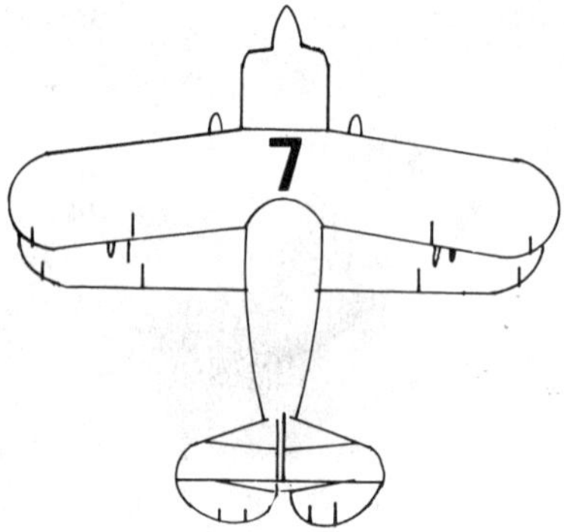

Covering the Fuselage

There are two methods of applying fabric to aircraft structures when using either linen or cotton. Regardless of size or shape there will be joined edges of the cloth. In the case of aircraft with V_{ne} (never exceed) speed of 150 mph or more there is no choice; seams are to be machine-sewn with approved quality thread and with all open ends hand-stitched (except for other specified means as per type). An airplane of less than 150 mph max. speed can be covered by the "blanket" method having over-lapped edges cemented to the frame. This rule applies to all parts of the airplane.

There need be no particular sequence in covering the parts of the airplane, but we will start with a description of the most practical way to cover the fuselage. Wings and the other surfaces have the cloth fitted and attached in basically the same manner, and for those, specific considerations will follow in Chapter 8. We recommend reading entirely through the book, then returning to its parts for guidance as you work, regardless of sequence used.

You may be fortunate in having the old fabric as a guide, if it has been carefully removed. It can be very helpful in details, but unless it is the original factory cover (increasingly rare now) be wary of being led astray, because a lot of cover jobs flying are not the best. But there is no harm in looking to see how it was done and you may see how *not* to do it; this too, is beneficial.

We make note here that many specific synthetic covering materials do not have sewn seams or other sewn edges. If used on ATC'd aircraft they are to be put on and finished as per *approved application methods* for named, specific types and models of aircraft. This repeated to avoid confusion or conflict regarding methods of attaching the covering and in the doping and finishing of it. For those coverings, follow the manufacturer's instructions for the specific aircraft. In general, the "blanket" method will apply and will serve as a guide; the envelope-type is used basically for pre-sewn cover kits.

For the nicest finish on any airplane the seam-sewn "envelope" method is the best because there is less doubled fabric in the overlapped cemented edges, and primarily because the fabric is free to draw up more evenly since it is fixed to the frame only at terminal edges; it is in one piece instead of in separate sections.

Measuring and Planning

Your fabric was purchased with consideration of the easiest and most economical layout for attachment to both wings and fuselage—the "big pieces." The fuselage often presents the most problems for it is almost always a very odd size and shape to wrap gracefully. The weave of the cloth must run straight with the slipstream. There will be a few areas, such as underneath the aft end where the side covering pieces come under and join on the bias, where this does not hold true, but they are negligible.

Measure and plan in detail before cutting; fabric is too expensive to waste. By cutting carefully, many of the remaining portions can be used elsewhere on smaller parts.

Pieces cut for the glue and overlap method can be cemented in place as they are cut (as done in the series of photos illustrating the method in the book), or they can all be cut first before starting to put them on; this would depend mostly on having a suitable place to lay or hang them in between time, as wrinkles must be avoided. For descriptive purposes, we will go step-by-step: cutting, then attaching, doping, etc.

Cut the two side pieces, allowing several inches extra beyond the amount needed to follow the contour of the surface. "Fuselage length" dimension is a straight linear measurement, and using that would leave you short on cloth. Take actual *contour* measurement. Use the wood, spring-type clothespins to hold the fabric to the upper longerons or perhaps other frame members (on cabin planes over-head formers may be used) which extend nearly the full length of the fuselage. The fabric should be one piece from firewall to tail-post unless aluminum is put on the forward portion. Where this top member is not straight with the thrust-line, keep the fabric straight and clamp it as necessary to continue fitting and cutting.

Allow an inch over top and bottom edges to wrap the fabric around these members. If the cover is to be lapped and glued, this portion, over and inside, would be cemented to the airframe; it can be done now if you wish, or lay carefully aside until all cutting is completed.

Before final cutting, clamp the fore and aft ends where they will be secured and readjust as necessary until the fabric hangs straight and even the full length. Use plenty of clothespins so there will be no appreciable sags. (We are assuming the work is stationary, rightside up). Let it hang this way, both sides clamped, while the top cover piece is fitted.

Note: If the sewn "envelope" method is being used, all pieces must be clamped, then pinned, in place. Pieces can be put on individually only when lapped and glued.

Cut the symmetrical top piece to cover the maximum width and length allowing for contours, as with the side sections. Do *not* cut a tapered section to an edge of the fabric—but centered in line with the weave—as that would result in a diagonal thread weave out of line with the airframe. To locate the center, fold the material in half lengthwise and mark the centerline at each end which can then easily be aligned on the airframe with any appropriate members fore and aft. With each end held by clamps, work along the adjoining side corners where top and side pieces meet. If it is to be lapped and glued, cut a clean, even edge allowing a 1" overlap along the top frame member.

Fitting The Material

Refer to the series of photos that illustrate this method of recovering the Fleet 16-B biplane fuselage. The rotatable front-end support greatly facilitated this particular job; it is much easier to fit and put on any material when the work can lay flat. Details to be looked for in this series will be mentioned as we go along, referring to Fig. 7-1, as example, showing the fuselage supported in the rotatable stand. Of course there has been much of the explanation already written which you will now find illustrated.

Please differentiate between description of sequences for cementing compared with sewing. We will go ahead with the lap and glue method shown and then return to cutting and fitting for the sewn envelope.

Figure 7-2 shows the linen being cut for the first side, rolled out on the side of the fuselage itself for direct measurement; a clean floor space would do, but since the fuselage could be easily turned

Fig. 7-1. Fleet fuselage ready for cover; supported at front end by special stand which allows frame to be rotated for covering and doping, so all work can be done right side up.

on its side, this is simpler. Notice the pliofilm floor-to-ceiling partitions for this home "paint and dope room."

In Fig. 7-3 the first (top) edge has been cut along the upper longeron and is held with clamps while cuts are carefully made for clamps to which cockpit-coaming will attach. The other (bottom) edge still hangs free.

Fig. 7-2. Laying out and cutting off the fabric for left side of fuselage. Alternate way is to roll it out on clean floor, measure and cut.

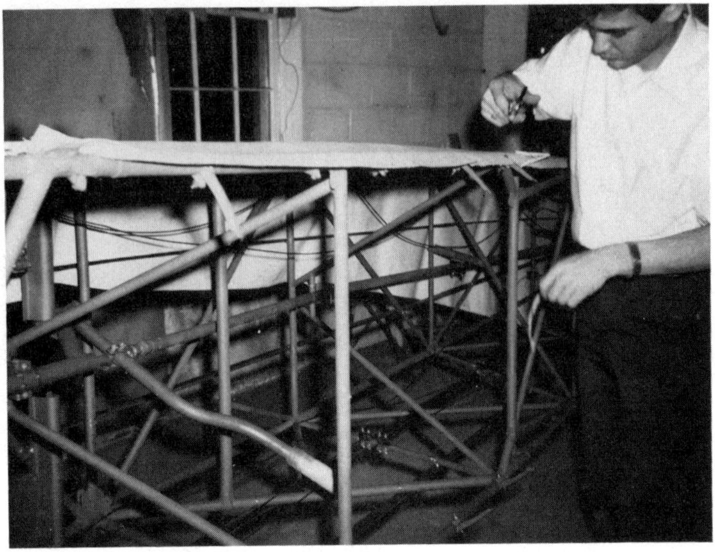

Fig. 7-3. Cutting carefully for fit around clamps and at cross-members. In this instance, linen was cemented to this top longeron before the bottom edge, which can be seen hanging, was cut off.

After the sides and the top are cut and fitted—they may or may not be cemented in place now, your choice—go underneath and consider the belly, which needs to be just as well done as every other part of the airplane. Unless your fabric is very wide and the fuselage very small, it will be necessary either to splice in a wedge-shaped piece covering the area back to where the fabric of the side panels will meet on a bias and be seamed along the centerline to the tail-post, or else a full bottom cover, similar to the top, will be overlapped and glued, or seamed, to the side fabric. The latter (full bottom piece) is used with the lap and glue cover. It is a choice with a sewn envelope. Follow the same procedure used for the top piece for a neat, symmetrical job. For a top-side-up fitting, the only way to hold the side edges along the bottom longerons is by pinning to the side fabric, which by now would be cut with a 1″ margin.

If seams are to be sewn for an envelope, replace the clamps with pins along all corners. Bring the edges together *outward*. Space the pins about 3″ apart and "stitch" them a couple times through the cloth so they will stay in place better during the fitting and handling while preparing to sew the seams. It may be necessary to re-pin from time to time to get all sections even and hanging straight. The name of the game is *patience and care*. Trim the fore and aft ends to their 1" allowance and reclamp them in place.

For an envelope. with all panels pinned along the seam-lines and trimmed to a 1" allowance, stand away and look at it. There should be no excessive sags nor folds and all should have weave in line with the slip-stream. It may look awfully loose but don't worry about that. The shrinkage of the fabric takes up an amazing amount of slack, and it is far better to start with slack than to pull it up too tight during the fitting. On a very light airframe, formers and even longerons (with widely-spaced cross-members) can be pulled inward by the tautening of the doped fabric if it has been pulled too tight by hand during this initial fitting. Only very flimsy structures will be deformed by the tautened fabric if it has been fitted properly, i.e., *evenly but not tight*.

Pencil mark along the seam-lines (pin-lines) on both pieces of the fabric the full length of each seam. Mark for any necessary cut-outs as a guide for machine-sewing. With pinking shears, cut away excess fabric, leaving about ⅝" allowance from the seam lines if this is to be an envelope cover. For lapped and cemented cover leave one inch allowance. Trim carefully, keeping a constant edge distance. See Fig. 7-4.

Fig. 7-4. Marking and cutting away excess along bottom longeron.

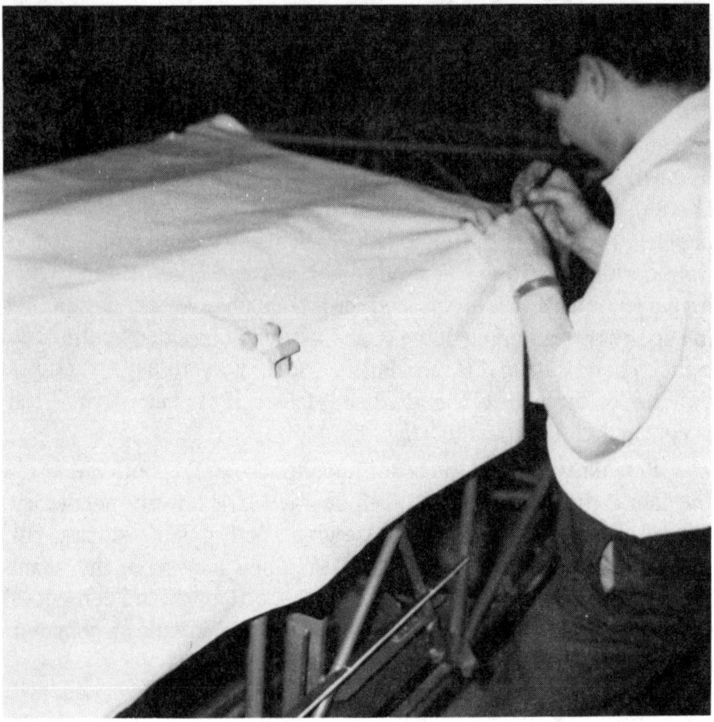

Fig. 7-5. Marking and cutting for gear fittings so fabric will lay even for good fit.

Leave at least an inch at the ends and at any other openings or edges for proper secure attachments as per individual aircraft requirements. Cut-outs for fittings, controls, etc., should be minimal and cut at this time only if the envelope can not be pulled over them (Fig. 7-5). Sometimes it is necessary to make a cut from the nearest edge to go around something that protrudes too much for the fabric to be fitted over (as with the ailerons, Chapter 5). In this case, *after* the fabric has been stitched and re-fitted, and *before* water-shrinking hand-stitch the cut as per *AC 43.13-1, Fig. 3-10.*

Sewing

If it is to be sewn, very carefully work the pinned envelope back off the fuselage. Keep the aft portion pulled clear to avoid bunching and tangling. Handle it gently to prevent pins coming out prematurely. And here, a good wife is your best help. If you don't have one, borrow one. Someone handy with a sewing machine is now number one priority.

Aircraft machine-sewing thread comes on big spools designed for commercial work. We rigged a piece of round wood stock (cut from a broom-handle) drilled inside to fit the sewing machine spindle inside the spool. A size #16 needle handles the aircraft machine thread. The bobbin must be wound with the thread too, of course.

Plenty of working space is needed. A long table would be fine but most of us must make do with a clean floor for the mass of cloth to lay on. *Never* allow any oil of any kind to contact aircraft fabric at any time; the fibers would absorb the oil, then not the dope. The work must be kept *clean* until the final finish is completed.

If your fuselage is exactly the same on both sides (cut-outs, etc.), seams can be stitched as pinned. If not, it must be reversed and stitched inside-out; re-pin with the matching pencilled seam lines together. This is the reason for marking *both* pieces along the pinned-lines while fitted on the frame initially. Stitch a little to the *outside* of this line to be certain the envelope will be big enough to go back on the airframe all the way. You might leave one bottom-to-side seam unstitched until you try it on for size. It could then be stitched a little to the inside or outside for best fit, if necessary. Because of the doubled material resulting at the inside of all corners, the fit will be a little more snug than you may have anticipated and this is the reason for keeping slightly to the outside of all seam lines.

After stitching and while the envelope is still wrong-side out (or inside-out), iron the seam edges flat so they will lay smoothly along the longerons or other corner members. This provides nice even surfaces over these corners without lumps or hollows. Iron out bad wrinkles and sharp fold-lines, particularly in linen. Otherwise, these will persist (in linen) even after water-shrinking and doping. *Do not wet or dampen before ironing.* You'd have pre-shrunk fabric and could never get it back on the airframe. Iron out wrinkles *dry*. And never stand a roll of fabric on end; sags and wrinkles form in it. Always lay a roll of fabric flat.

Turn the stitched envelope right side out then pull it on gently and carefully. The support under the aft end must be shifted a few times until the fabric is worked forward of it, then pad the support so it will not chafe through the unprotected envelope. If the tailpost can be supported clear of the fabric, that would be helpful. An iron cross-bar was slipped into the open tail-post of the Fleet illustrated. The work can be suspended from rafters, etc., but it is easier to work with the airframe on solid supports; hanging a wing can work pretty well. Be patient in working fabric over any projections. Keep it straight and even all round. When it is completely on and adjusted, pull firmly but not hard from the front end and re-clamp it there until the hand-stitching is done, permanently securing the open end and other loose edges. The PT series Stearmans have all of these edges fastened with aluminum strips held in place with small P-K screws.

Gluing

Let's return again to the lapped and glued coverings, to bring both styles to the equivalent condition, ready for water-shrinking.

Before starting to cement any fabric to the airframe, have all materials and equipment listed close at hand. If there is any concern about the surface (finish) of the floor, walls or furnishings within the work area, cover them with something heavy enough to prevent dope going through on to any varnished or painted finishes. It is advisable to cover the floor with several layers of newspapers and then a canvas over the papers. There are always dribbles. Have an extra, small container with thinner for the small brush that is used for applying the lacquer cement, to keep it clean and pliable.

Neatly fold and set aside all but the one piece to be worked with. Start with the sides, just as you did with the first fitting and again use the clamps to hold the fabric in place, removing them only far enough ahead to be able to lay back the fabric to expose the frame member so cement may be brushed on. Put it over the top and around the *inside* edge. *Never allow cement or dope to drip onto the inner surface of the cloth.*

The accompanying photos (Figs. 7-6 through 7-14) illustrate putting the linen on the Fleet fuselage: left side, right side, bottom and top. Note the careful fitting around clamps and cementing the linen to the *inner* surface of the tubing, which leaves the outer part of the cover free to tauten evenly; next, the many clamps to be worked around at the aft end. Rudder control cables and tail-light wire have been taped out of the way until dope and fabric work is done. After linen has been cemented to the longerons and tail-post, it is finished off at the front former; metal extends on forward to the firewall. Then, start on the top edge of the right side cover; note papers inside to catch drips so they will not spoil the inside of left cover (Fig. 7-10). Next, the bottom cover piece (Fig. 7-11) is fitted and trimmed, with special care around the landing gear attach fittings. Notice that at no time has the linen been pulled; it was layed in place, clamped to keep it from shifting, trimmed and then

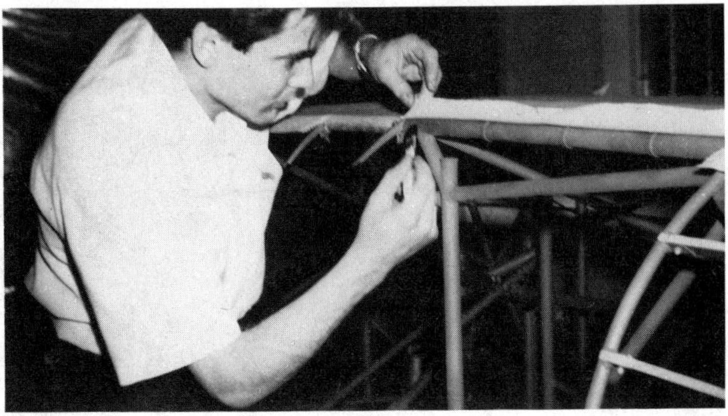

Fig. 7-6. Cementing edge of linen to top longeron.

cemented. This piece (bottom) was first cemented along the front former so it would lay evenly and straight along each side, progressing aft.

Note: The change-over here from the lacquer cement to dope. The former is used for cementing fabric directly to anything other than fabric, but for fabric-to-fabric, use *dope.* Brush the dope only where it is needed for the overlap, sufficient to hold the pinked edges, because once in the fibers of the fabric, it is there for good and it would always show a slight depression. Never allow it beyond where the finishing pinked tapes will fully cover. This point will be clearly shown in a following picture (Fig. 7-16). As soon as the outer fabric is firmly stuck to the one underneath, dope over it

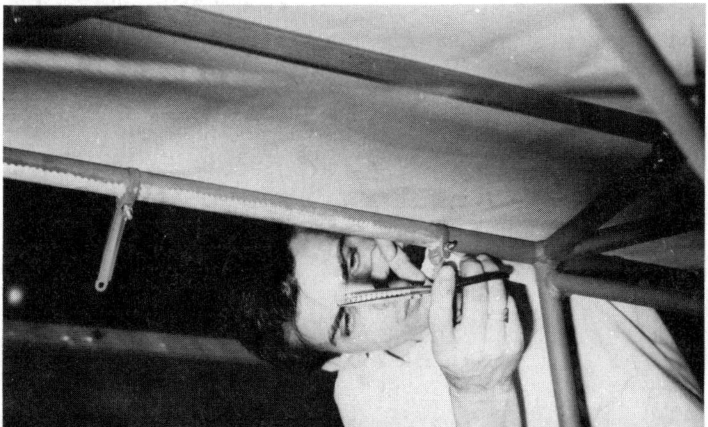

Fig. 7-7. The lacquer cement is applied to the *inner* edge of tubing. The fabric is *not* pulled while being cemented in place.

Fig. 7-8. Careful trimming and fitting at aft end.

then go on to the next section. About 18″ to 24″ is about all that can be done well at a stretch because good adhesion is essential and it must be done carefully and thoroughly.

So, first the side pieces, then the bottom, then the top, making certain always that all pinked edges are cut straight and even. Do not pull the fabric tight; let it lay or hang easy (Fig. 7-14). Fabric that is pulled tighter one way or the other will show ripples after tautening. If there is any excess of cloth at the ends, distribute it as equitably as possible as the ends are cemented to the frame (or stitched, as the case may be). Shrinkage will remove wrinkles amazingly well. Follow the same procedure in closing any other openings. If the cover has been carefully fitted and put on there will be very few sags or wrinkles.

Hand-stitch any cuts that were necessary to clear projecting parts, and cut the smallest openings required for any other fixed projections. The water-shrinking which follows will enlarge these cuts and holes and it is nearly impossible to anticipate their final precise location, but they will be finished with pinked reinforcing tape or patches later. Always make all cuts with pinking shears and keep them as small as practicable. However, the cloth must rest free of all projections so that it is not bunched and wrinkled.

Water Shrinking

Most tap-water contains minerals which could shorten the life of the fabric or reduce adhesion of the dope so it is advisable to use only bottled distilled or mineral-free water for shrinking. It is sold at most drug or grocery stores. Four gallons will be plenty for an average size airplane. It is not wiped on but *rubbed in* thoroughly. See Figs. 7-15 and 7-16, where the rapidity of shrinkage is apparent. A turkish towel or washcloth is fine; any cloth that will hold a good volume of water for full and uniform wetting is okay. Work as symmetrically as possible so the cover will tighten evenly and not pull more to one side than the other. Especially with an envelope cover, keep the sewn seams where they belong, straight with the corners, along the longerons, etc.

In Fig. 7-16 the doped overlaps are clearly seen; water does not shrink them, but the doping to follow will. Figure 7-17 shows

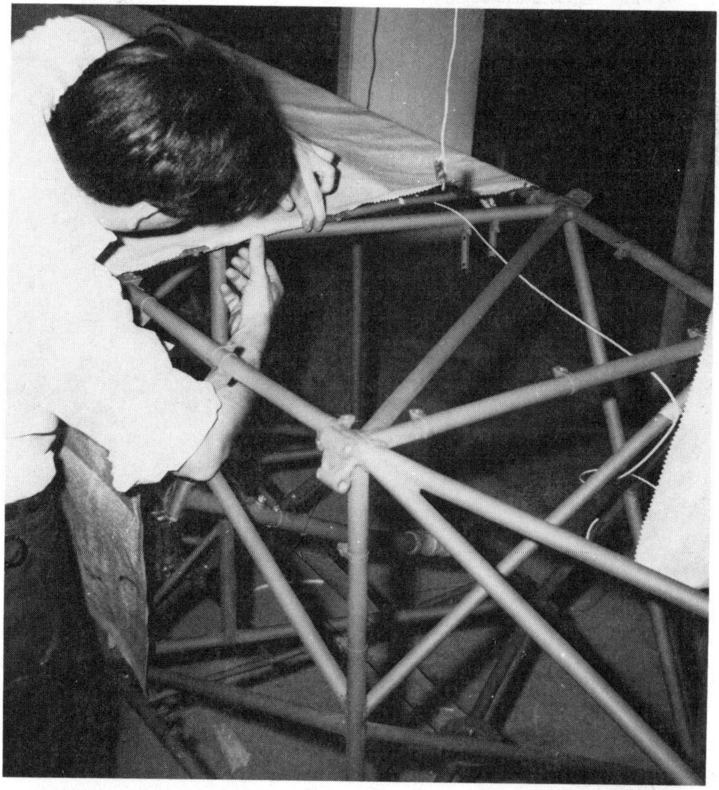

Fig. 7-9. Securing linen to the front (side) former. Metal coaming goes on forward to the fire-wall.

Fig. 7-10. Trimming at fittings and cross-members for cementing top edge of the right (second) side of fuselage cover piece. Note papers inside to catch drips, protecting inside of opposite cover.

the writer's homebuilt fuselage (Grade-A cotton) after water-shrinking, ready for the first, fungicide, coat of dope.

If synthetic material is being used, a hot iron pressed evenly and uniformly over the surface brings the cover to a desired

Fig. 7-11. The bottom cover piece is fitted and trimmed. Overlapping edges will be doped to side covers.

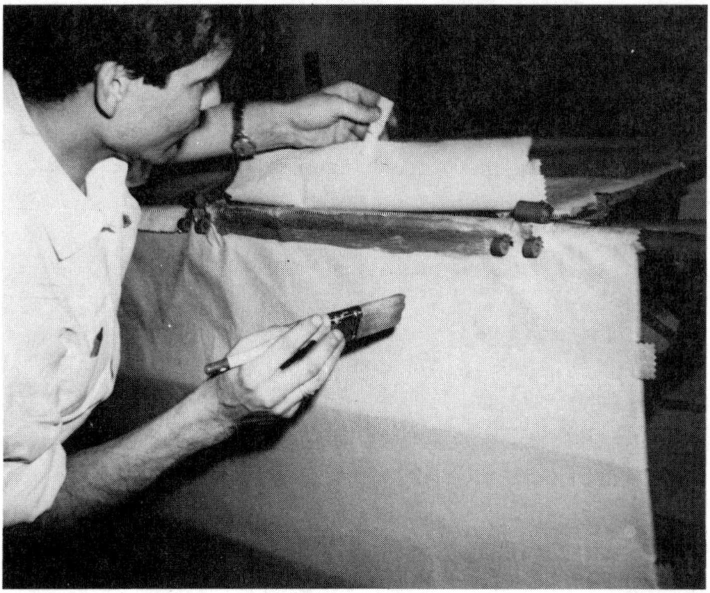

Fig. 7-12. Dope is brushed *only* on actual overlapped surface (bottom edge to side piece, on each side). Note that the front edge of this bottom piece was cemented to the front former before doping the bottom/side edges to better secure it, and keep it from shifting from side to side.

tautness. Do not dwell too long in any one place; keep the iron moving. A *moderately* hot iron does best.

Note: Due to the many kinds of materials and finishes of the synthetic type currently on the market, we can only recommend that anyone using them follow the instructions published for their

Fig. 7-13. Brushing dope over the lap, with care *not* to let dope go beyond the actual overlap area.

Fig. 7-14. The fuselage cover is all on. It is now ready for water shrinking. It has *not been pulled* at any time during fitting and cementing.

specific methods of application. However, except for doping, the following general information will still apply for all covering work. It must be handled carefully to avoid excessive ravelling and fraying of all raw edges. Special, matching sewing thread and rib-stitch cord must be used, and sometimes the finishing tapes also, usually so, in fact. So read this only as a guide for your *general information*, and follow the directions accompanying *your* purchased materials.

The most difficult part of covering the fuselage is now behind you. Take a day off while it completely dries with a DO NOT TOUCH sign on it. Then lightly wipe off any loose lint that may

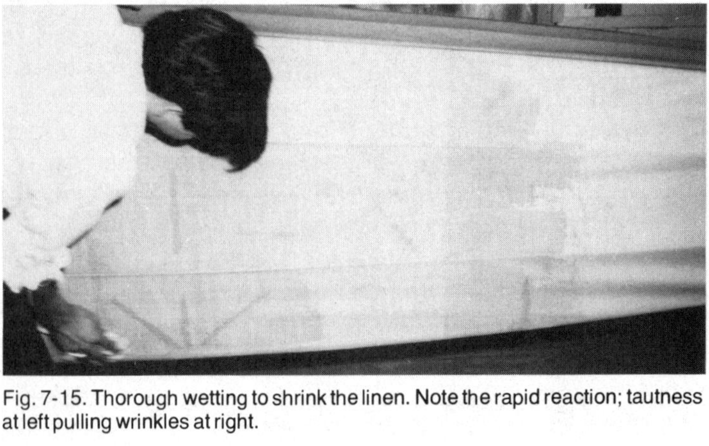

Fig. 7-15. Thorough wetting to shrink the linen. Note the rapid reaction; tautness at left pulling wrinkles at right.

Fig. 7-16. Completely wetted fuselage, uniformly taut. White lines are the doped overlaps. You can see why they should be at minimum.

have rubbed up. If it is a large fuselage, the cover should be secured with a few knots to underlying members to prevent vibration and wear, as described for inside the cabin. This is not generally necessary when there are stringers underneath as they keep the fabric stretched in a convex form and they are pretty close together. The ties are needed only where there are large flat areas.

Doping

If you have the facilities and materials (odd pieces of 2″ × 4″ will do) to make a rotatable "barbecue" stand, as illustrated, it does make the work much easier. Brushing dope upside-down is a messy business, and putting the tapes on evenly and straight is difficult; with help, a fuselage can be turned if not too large and heavy, but this is slower and it requires two people.

Temperature for working with butyrate dope must be up to 72°F and more if humidity is high (when it often helps to raise the temperature), but there is naturally a limit reached this way. In some places the humidity can be impossibly high and this is one of the primary reasons many people choose to cover with synthetics, though if they use butyrate there is no advantage in this respect. If the weather is warm enough to work with the windows open, that is best. But avoid drafts which can cause too fast surface drying with resulting "blushing." If you must work in a heated room, have some fresh air and ventilation, but no drafts. If the humidity is high—a seasonal problem in many parts of the country—raising the temperature helps and a de-humidifier is very effective if the work area can be closed off (see photos). The reason for this concern

Fig. 7-17. Homebuilt fuselage with water-shrunk Grade-A cotton, ready for first (fungicide) coat of butyrate dope.

when working with butyrate is that the dope will "blush" if it does not dry uniformly through its full depth. The surface turns a skim-milk, bluish, foggy white, almost iridescent. (See photo of Fleet rudder in Chapter 10.) Clear dope should dry *clear* and colors should be true. Blushing is not only a blemish, but the dope does not have its proper lasting nor adhesive qualities. So dope only when and where conditions are favorable. This can cause maddening delays, but it pays to wait.

Retarder can be added to dope to slow (retard) the drying and thus reduce the chances of blushing, but it is rather expensive and it is generally used only as a last resort, especially for quick repairs or in refinishing old work where it will aid in penetrating old dope because it slows drying.

Fungicide is mixed with the initial coat of dope to inhibit mildew and mold growth which forms on the *inner* surface of the fabric and would cause it to rot under the dope, shortening its life appreciably. Humidity and condensation in a closed hangar will totally wet all surfaces of anything in it, so although a plane is sheltered from the elements, moisture is everywhere, except in the very arid parts of the country.

The first coat of dope containing the fungicide must be thinned 50/50, plus the fungicide added as described in Chapter 6. Unless you are covering an unusually large airplane, one full amount of the "recipe", 1 lb. fungicide to 4 gal. of dope/thinner, will cover the whole thing. Mix it well in a wide-mouth 5 gal. can with a tight fitting lid which will seal it between periods of use. Because the fungicide tends to settle rather fast like paint pigment, stir it frequently during applications.

Use the large dope brush (Fig. 7-18). It will carry more dope for faster fill and build-up, and of course, cover more area per swath. Each coat should be as heavy as possible without getting runs or puddles. However, the fungicide coat *is* applied differently: it is *brushed in*, not just *on*, as with plain dope. It is to be brushed in thoroughly and evenly, completely wetting the fabric, but do not permit runs to develop, especially on the inside.

Extra ventilation while applying the fungicide is recommended. It does not have a particularly disagreeable odor, but by the time you have gone over a fuselage or a wing it really gets to you in a closed area. Have a window open somewhere, preferably with a fan to exhaust the dopey air. Dope fumes, like gasoline vapors, seem to accumulate at the lower levels, so hand-doping the bottom of a fuselage (if it can't be turned over) can be something of an ordeal without good ventilation. This is particularly true with this first coat, not only because of the fungicide, but also because the application is slower since it must be carefully brushed in.

Always take special care when brush-doping the sides (or any vertical surface) to avoid runs inside caused by brushing runny, thinned dope too much in local spots, and dope running inside when working around a hole or cut-out. It helps to put paper inside such openings to catch possible drips. Any internal drips or runs result in permanent blemishes. So go very carefully around all openings *at all times*.

Fig. 7-18. Brushing in the first coat of butyrate mixed with fungicide.

Try to plan ahead for the doping. Do not allow too much time (days or weeks) between coats of dope, which would permit it to harden layer by layer instead of achieving through penetration with each coat. Any time the work must be left, cover it with clean paper or cloth; old sheets are fine for this purpose.

The following coats of clear dope *can* be sprayed but brushing is preferred because it better penetrates the fibers and builds up faster. It is the absorption by the filters that makes a good finish adhere well. Synthetic fabrics *do not absorb* and as a consequence the finish will sometimes crack and peel.

Brush coats of clear dope should be cut with just enough thinner to let it spread evenly without "roping" or pulling, which would result in a rough surface and uneven cover and penetration. A mix of approximately 10% thinner, by volume, generally works out best. A thinner mixture is necessary for spraying, which is one reason brush doping builds up faster and better as base for the silver and finish. If the dope is too thin it will not tauten up well either. It is the "solids" in the dope that gets the job done; thinner just evaporates. Solids are determined by the chemical structure; do not confuse the term with viscosity. Brush with smooth, even strokes, as full as possible without runs or puddles. The object is to keep the application uniform while getting the best build-up per coat. Do not "scrub" with back and forth motions; lay it on and spread it out. Overlap strokes for even coverage.

After the second coat (first clear) you will wonder what awful thing you did wrong—the fabric will be alarmingly slack and baggy all over. Don't worry. If you did this on the interior you will have been through it before, but it is a shock to a beginner. As the third coat dries you will notice it starting to draw up even again, and it gets better and better as each subsequent coat goes on and the weave fills and a gloss shows.

Following the third coat, a first sanding of the entire cover should be done *before* pinked tapes are put on so there will be no roughness under them, and before dope builds up on the nap of the fabric. Use #180 grit emery paper, and go very lightly over corners and covered projections (Fig. 7-19). It is very easy to cut through the cloth fibers where there's something solid under them, particularly any kind of point or edge. And *any hole requires a patch.*

Taping

All pinked tapes and reinforcing patches should be put on as soon as the fabric starts to tauten; it should be ready following the third coat. They will not lay on smoothly and evenly if the fabric is still too saggy. They should go on as soon as possible so the total build-up of dope will be uniform. The standard 2″ and 3″ tape and the reinforcing patches are best applied using the wide brush so the dope will extend beyond the edges of the tape.

Figures 7-20 through 7-26 illustrate doping on the pinked tape. A rather heavy coat of dope should be brushed on as great a distance as will stay wet while the tape is layed on and brushed down (Fig. 7-21). Go over this section immediately with another brushful of dope to wet it and lay it completely. It helps to brush dope on the *underside* of the tape as you lay it (Fig. 7-22). Be certain every bit of the tape is in contact with the cover. Rub out any air bubbles with the fingers (Fig. 7-23), if necessary, but continue laying the full length of each piece of tape as quickly as possible rather than concentrating overly long on one or two local spots; then go back and work them out *before* the dope dries. If you stop at some point long enough for the dope to dry it will leave a line across the tape. Get it all on, well wetted with dope, then work on the local hollows until the tape or patch is completely stuck down to the cover. Do not pull the tape, as this makes it very difficult to persuade it to go into the hollows. Sometimes the pinked edges persist in turning up; they must be worked down smoothly.

Pinked tape should be used over every place where the cover lays in contact with any part of underlying structure or wherever it *could* contact it with vibration. By the third coat of dope these places can be determined by light finger pressure. If only a slight

Fig. 7-19. First sanding, following third coat—very easy over all edges.

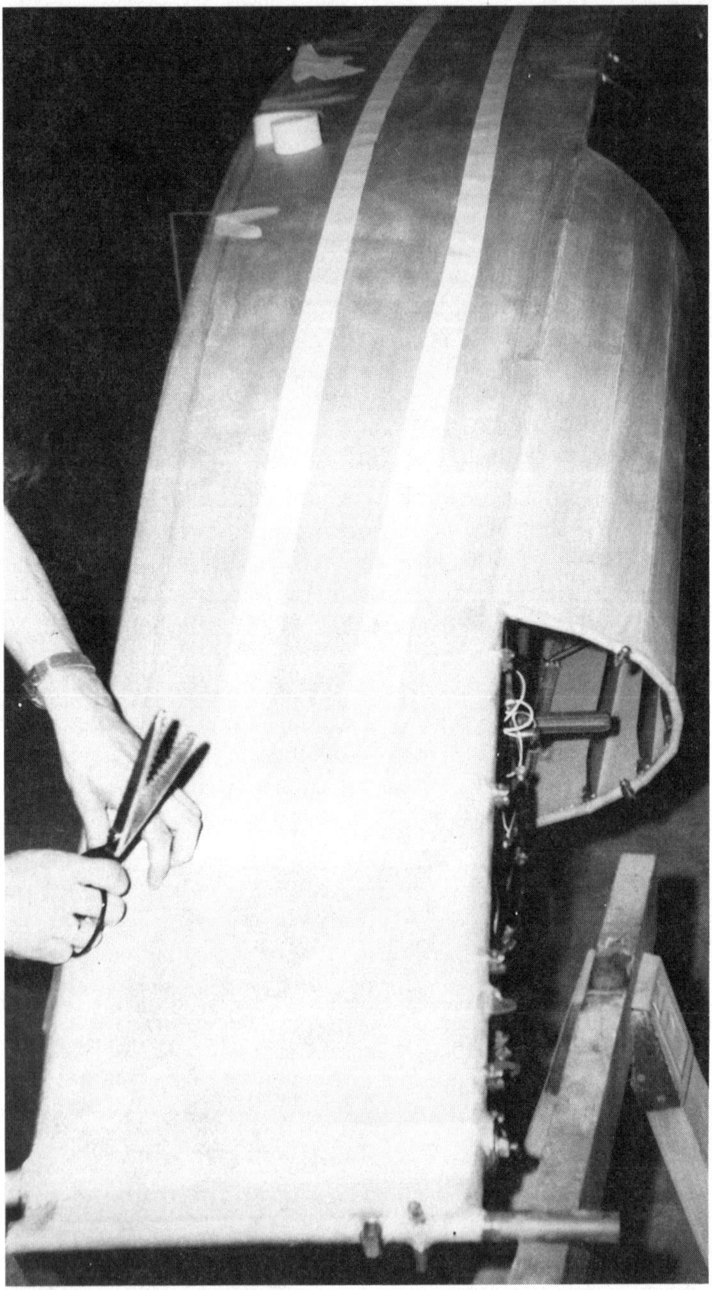

Fig. 7-20. Laying out the pinked tapes for the left side. Right side will be duplicate. Tapes go on after the first sanding.

indentation brings the cover in touch with something (*anything*) underneath, put tape over it. If it is just a local spot, cut a neat circular or oval patch for it. Full length strips of tape are to be used over all corners, stringers, seams and closures.

The lengthwise tapes go on *before* the finish tapes at the fore and aft ends; all others, first. Cut and fit neatly at all projections and fittings. See Fig. 7-24 showing careful trimming for the gear attach fitting, etc.

Make reinforcing patches for around openings for control rods and cables of old duck or twill (worn out pants come to a worthy end this way at our house); iron to remove wrinkles, for they are already pre-shrunk, and be *certain* of their precise locations. This is where the old fabric is of real value. It may be necessary to hold control surfaces in their respective relative positions to project the route of control cables in their *rigged* positions to determine correct location of the holes for cables to exit the fuselage cover. It is so easy to be mistaken when trying to locate these openings and

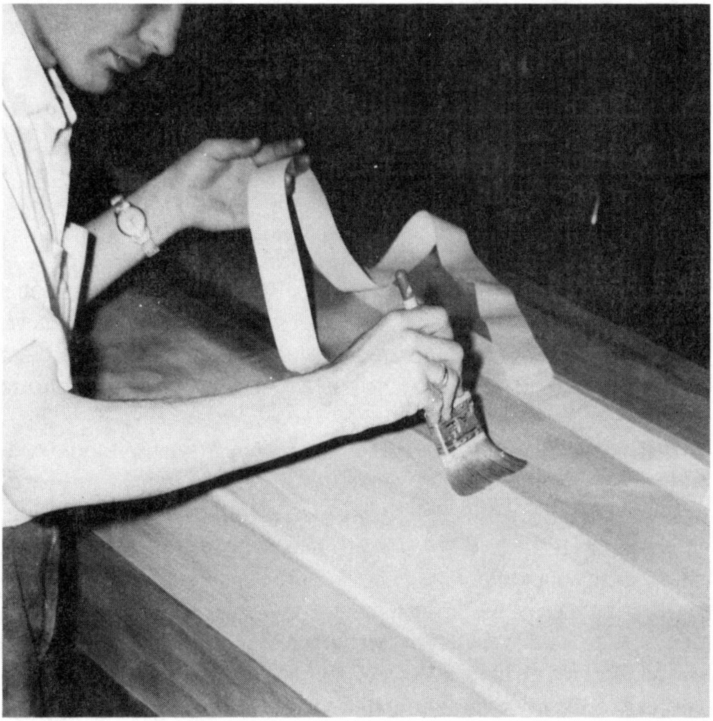

Fig. 7-21. A heavy coat of dope is brushed on as great a distance as will stay wet.

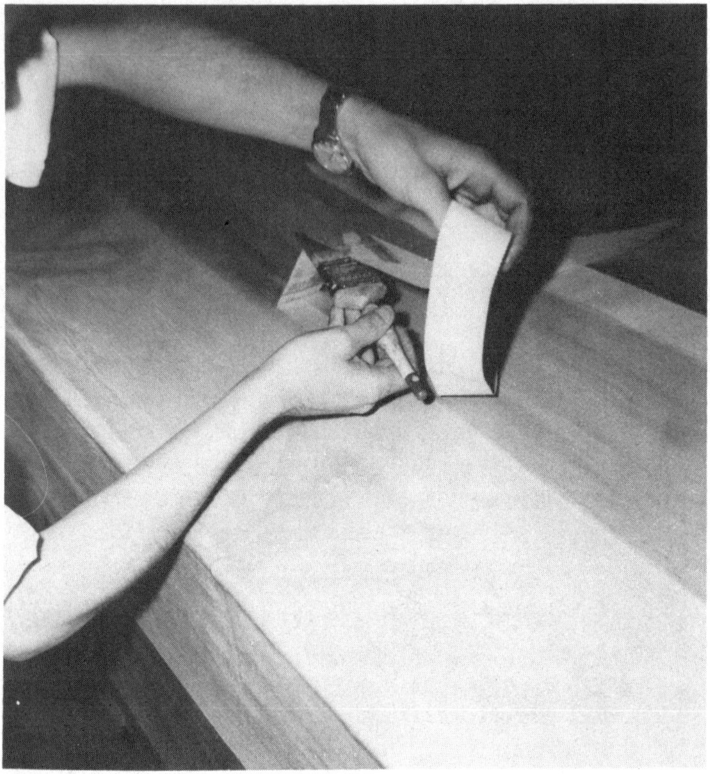

Fig. 7-22. It helps to brush dope on the underside of the tape as you lay it.

then when the plane is assembled and the controls connected, things don't match up. So more holes and fresh patches on the new cover. The tear-drop shaped patches for the Fleet rudder cables were located by reference to the old fabric. The pair on the bottom are for the tail-wheel steering cables (Fig. 7-25).

Re-dope all tapes and reinforcing patches to tauten them and to bring up the fill of their weave to the equivalent of the base cover. Sand the pinked edges particularly well to "feather" them to the base fabric (Fig. 7-26). Do not make the actual cuts or holes until assembly and rigging.

There is no need for the fungicide on the tapes.

The cover should not be permitted to stick *un*intentionally to any support for stringers, control cables near the surface, etc. The purpose of "the plumber's friend" — the rubber suction cup — is to pull the fabric free from any such under-lying member. Do not use it in wet dope, but after the surface is dry to the touch. As it dries completely it will tighten up and remain clear of these internal parts. To allow dope-soaked fabric to lay on wood or metal parts could result in the eventual lifting of varnish from wood or zinc chromate from aluminum or steel, and besides, the fabric is not free to tighten evenly. To lift stuck fabric with a pin leaves holes that never close and fill because the fabric tautens *away* from the pin-holes. Some of these things we learn the hard way.

There is really little difference in working with linen or cotton; linen is heavier and stiffer and it takes a little more dope to fill and more sanding, because linen is by nature a rougher cloth. But a good linen cover will last at least double the life of an equally well done cotton cover. If weight is not of too much importance, linen is first choice. Whatever the fabric used, it takes a *gloss* base of clear dope (or equivalent for synthetics) for a really good final finish. It takes a full six coats of clear to give a gloss, with sanding after alternate coats.

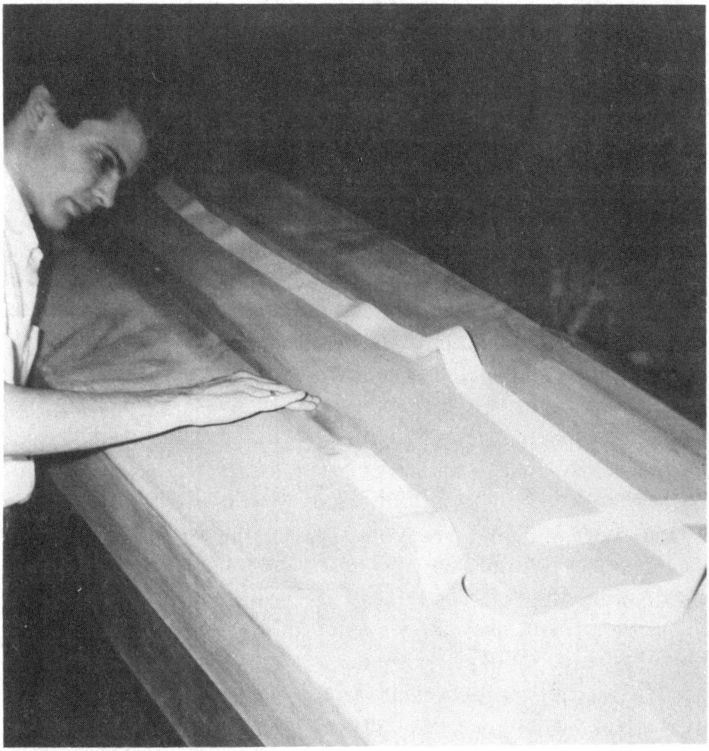

Fig. 7-23. Rub out air bubbles with the fingers (*clean*, or course), then brush over with dope and continue laying the tape.

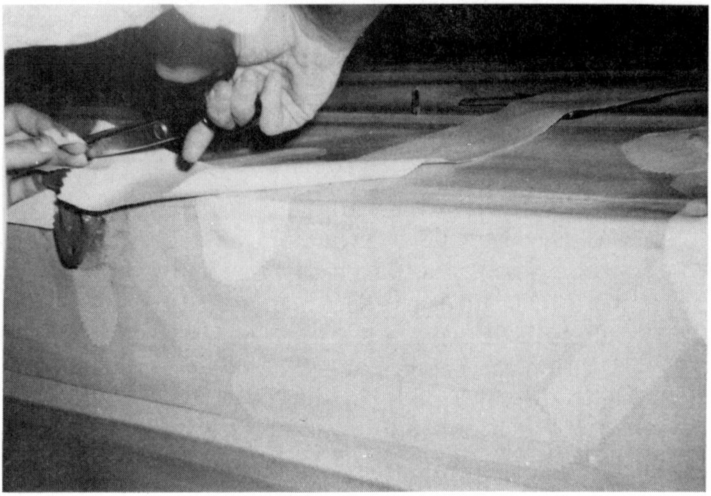

Fig. 7-24. Cut and trim very carefully for clean, snug finish around all protruding parts. Tape is used over all underlying structural parts which fabric does or could contact and chafe.

Too much dope can be put on. A solid glaze is not desirable. The weave should show, but the surface should be smooth to the touch with a satiny appearance. Excess dope is an unnecessary expense, it adds weight, it can harden and crack, it does continue to draw up to some extent in warm, dry air and could therefore warp a very light structure.

Light hand-sanding overall about every other coat will give a silky-smooth surface; if the clear is not smooth, the finish will not be. Give special attention to all pinked edges. Be careful not to over-sand the adjacent fabric nor over edges (corners). The pinked tape is no tougher than the cover; it is Grade-A cotton. Many old planes had leather reinforcing patches around the openings for controls, etc., but when stitched on, their edges would curl outward and be rough. We like duck or twill better. They can be doped down tightly with their pinked edges "feathered" by sanding; dope will fill their weave and the finish color will remain the same as the rest of the cover for a neater appearance. The elongated, tear-drop shape allows for the opening to be extended if necessary to accommodate the action of the cable or rod.

There must be provision for drainage and ventilation of *all* aircraft components. Unless there is an opening of some kind at the low point of the bottom cover (with plane sitting at rest as parked) put one or two seaplane type grommets at the lowest point. It is better not to cover over an open-ended tail-post. Leave it open, or if covered, put a seaplane grommet under it. Moisture can accumulate down in there and eventually fail it from rusting. The lower aft end is a common location of severe rusting of both longerons and tail-post. Remember: holes for seaplane grommets are cut *before* the grommets are cemented on (before final coats of clear dope).

Depending on the particular airplane, inspection holes may be required in the fuselage, and perhaps access holes for controls for assembly purposes. The plastic reinforcement rings and related "patches" should also be put on shortly after the tapes have tautened as they most often overlap the tapes and they must have a smooth, flat surface to adhere to. Following coats of dope will help seal and secure them. The cellulose rings and grommets are soluble in butyrate, so do not brush dope on the raised open portion of the seaplane grommets or they will soften and collapse. Holes in plain flat grommets and inspection hole reinforcing rings are cut out after all doping (color included) is completed. All holes are to be cut very carefully with a very sharp knife or razor blade — be careful it doesn't slip and cut the fabric! *Never punch in drain holes.* Whatever the cover may be, its fibers would be pushed up inside all around the inner edge of the hole which would result in moisture and dirt being held in instead of drained out. Make a clean cut around any hole.

Fig. 7-25. Pinked tapes all on. The "tear-drop" patches on bottom are for tail-wheel steering cables; those on each side are for rudder control cables. Note bottom end of tailpost (right fore-ground) has been left open for drainage.

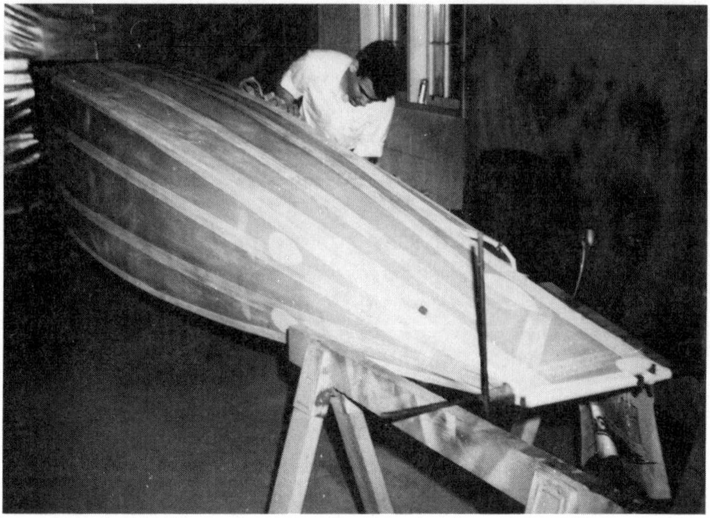

Fig. 7-26. Tapes and reinforcing patches are all sanded with special attention to feathering down pinked edges. There will be at least three more full coats of clear dope, alternated with sanding.

A further note about seaplane grommets: the opening faces aft. toward the rear. Airflow past the opening acts as a venturi, drawing air, moisture and dirt out of the interior of the structure. They are very effective and well worth the few cents extra cost. Since the opening is at the rear, dirt and water is not blown nor splashed *into* them, as can and does happen with plain grommet holes.

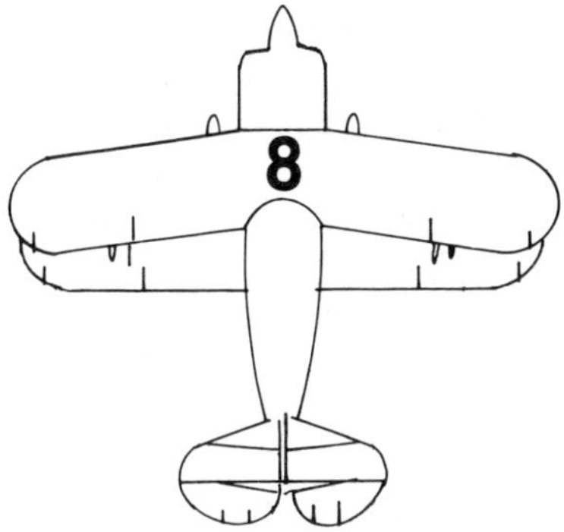

Wings And Tail Surfaces

We discussed the covering of the fuselage in the previous chapter though there is no set reason to follow any prescribed order. Very often it is the choice of someone new to dope and fabric work to do a small surface first to get the feel of handling the materials. In case of some "disaster" there is little loss of material and time and one can find out where his error was—although the point of this book is to help you do the job *without* having something go wrong. But a small surface is much less complicated to work on and it can be hung out of the way while the rest of the aircraft is brought up through clear dope and all parts are ready for silver.

If there must be a long period of time between the covering of one part and the next, it would be better to go ahead with application of silver dope before the clear dope hardens. The constant objective in getting a long lasting finish is good penetration, and therefore, the more closely one coat follows another, the better. However, there is a quite reasonable latitude; time is not so critical as to rush any part of the work. The sequence is your option and is most frequently determined by your facilities and the weather conditions.

Once the cover is on any part, and at any stage, protect it with a sheet or other light, clean cloth from light, dirt and moisture.

The procedures given for covering the fuselage apply also to all of the surfaces and the wings, but since the wings require additional attention we will go into detail regarding them.

As with the fuselage, the V_{ne} speed determines the method of covering them and certain requirements are very explicit. Refer again to FAA *AC 43.13-1* or the older *Manual 18*, which describes the procedures to be used—how the work is to be, rather than how to do it. The manual text and diagrams are not repeated here, but they are referred to for your guidance and they are essential to the work. *AC 43.13-1* figure numbers will be given in italics.

Whether blanket (glue and overlap) or envelope (sewn) method is used, the lengths of the fabric must be measured out and cut, then edge-stitched together to form one piece large enough to cover the entire wing completely (both sides). Do not simply double the chord length; the curve of the airfoil section adds considerably to this sum. Take an actual measurement around the longest rib (if a tapered wing). Measure around the longest rib in the aileron well. One or more lengths will be shorter in the aileron well area; this is true also if the wings have flaps. The relative pieces must be planned and cut to length to insure covering the greatest distance around the particular portion of the wing. This will vary according to plan-form, of course. Be sure to allow for the over-lapping of the seams in accounting for the overall span length measured. There should be sufficient fabric to cover the root end of the wing and at least one inch in excess at the tip, for either method used.

Measuring

Clear and clean sufficient floor space to roll the fabric out flat for accurate measurement and cutting. Allow 4″ to 6″ extra at both ends for the fitting, pinning, seam overlaps and/or minor goofs. Extra material can always be cut off, "not quite enough" can spoil your whole day. Mark off the measured lengths at both edges of the cloth and draw a line across to follow for cutting. It is best to number or otherwise identify each piece for its respective position for edge-joining correctly. When you have cut four, five or maybe more lengths of fabric, it would be very annoying to have to remeasure every one again to determine which adjoins which along what edge. Plan ahead.

Fig. 3.3 Manual AC 43.13-1 illustrates seam stitching and the methods approved for attaching the cover at open ends or at the root end by tacking or sewing. Note that a plain overlapped seam with double row stitching ⅜″ apart is approved for pinked edges, or selvedge edges with ½″ over-lap. Try to maintain even spacing and a 1/16″ edge allowance, for nice straight seams. Because of the great bulk of material that accumulates with each additional seam, a second person to help handle it is almost a must, so don't go dashing off to work with a: "Hey, Honey, will you stitch these together today?" It is a job for two. Of course, each seam must be pinned together before stitching. The 1/16" edge clearance is important to preclude any pulling apart when the fabric becomes taut during or after doping. The straighter and more uniform the seams, the better for both strength and appearance.

Can you imagine doing all of this for four wing panels for a "Jenny" or a Boeing 540-B or a D.H.-4? Many were done, and even now there is an occasional restoration, "as per original."

Much of the fitting procedure described for the fuselage also applies to the other components. Depending on the type of aileron fittings and the place you are working in, you may be able to suspend the wing with the leading edge down, which can be an easier position to work with. The fabric can be brought up around on both sides and pinned along the trailing edge so that it will hang evenly over both surfaces. If the wing or other surface is fitted while resting on sawhorses or bench, check frequently to be certain the fabric is straight and even on the underside.

Adjust the edge-stitched material to avoid seams running over ribs so: 1) rib-stitching will not go through seams, and 2) there will *not* be double thickness over ribs. This was one of the considerations in selecting the width of the fabric before buying it.

While pinning the fabric along the trailing edge and around the tip, be careful not to pull tightly in any one direction or area. Remember that the slack on the underside will have to draw up by shrinking. If the upper surface is pulled tight during the fitting there could be problems later. If the cover is the envelope type, the seam would be pulled around onto the side which has already been pulled tight, thus displacing the seam from the actual trailing edge where it belongs.

Leave sufficient material at each end of the aileron well to cover the adjacent rib surfaces, and also the root end of the wing. If this is not feasible, the main cover fabric can be brought over with a one-inch lap cemented down (after notching to follow contour), then a separate piece of cloth, cut to airfoil shape can be doped over this, as was done with the ailerons illustrated.

Do not cover over wood spar ends. They must be left clear for inspection purposes. Keep them clean. Put masking tape on them to protect from dope and leave them covered until the wing is ready to be assembled to the airplane.

For wing panels with ailerons and flaps, only the tips can be machine-stitched, then after trimming and ironing down the seam edges, re-fit to the wing for hand-stitching of the balance, i.e., aileron (and flap) well, remainder of the trailing edge and root end. See *AC 43.13-1*.

Gluing

For "blanket" (lap and glue) covering after the trailing edge has been pinned together (top and bottom pieces), pin around the tip, working both sides out smoothly and re-pinning as necessary to get an even contour on both sides, then cut away excess leaving 1″ allowance all around. Notch the allowance at close enough intervals to eliminate folds and doubling of material as it follows the contour. Remove all pins from the tip area and cement the bottom fabric around the outside of the tip, laying the notched tabs uniformly for even distribution. Dope over this the fabric from the opposite side. Pull only enough to draw the material smoothly around the contour of the tip. If the wing panel does not have aileron or flap cut-out, first close the trailing edge, then the root end, then the tip.

Many of these described details are illustrated in the series of photos taken during the re-covering of the Fleet ailerons; see Figs. 8-1 through 8-10. The first five photos show the fitting, trimming and cementing of the linen on the ailerons, to readiness for the fungicide coat of dope.

The attachment of the cover within the aileron and flap well should be done after the trailing edge portion is closed, and the means of securing if differs with various aircraft and types of structure. Follow any specific instructions available for your particular airplane, or refer again to *AC 43-31-1, Fig. 3.3,* if your cover is stitched. Do *not* depend on cement and dope to secure fabric with a *concave* well of ailerons and flaps. Any concave surface must have positive means of securing the fabric. Stitching is more reliable than tacking in wood (tacks can back out and/or split the wood) because the shrinking of the fabric tends to pull out the tacks. Screws with flat washers are used in some metal wing structures where the metal would chafe through stitching thread.

Sometimes it is necessary to cut the fabric to fit it over a projecting part; in such instances, make the cut to lay at the *base*, where the cover *will be* when it is in its correct place. If it is unavoidable to have an extensive cut, stitch it together after the fabric is in place, *before* water-shrinking. Do not worry about small gaps at openings for all of these places will have finishing tapes or reinforcing patches.

Fig. 8-1. The 42″ linen covers ailerons spanwise; the selvedge has been trimmed off and the bottom part of cover comes around the trailing edge and is cemented along the upper surface of it.

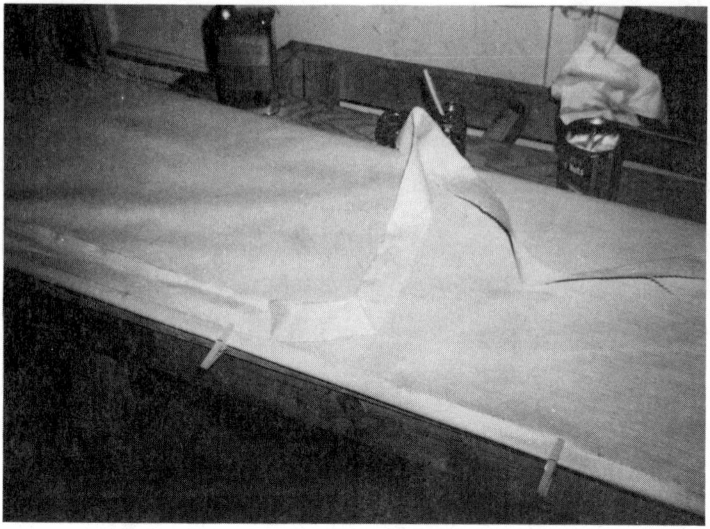

Fig. 8-2. The upper aileron cover goes around the trailing edge to overlap the bottom cover, to which it will be doped.

Fig. 8-3. The linen is cut to within an inch overhang and notched to follow contour of the rib (same each end for these). This is cemented down and another piece, cut out to rib contour (seen on bench in front of the cans) is doped on, overlapping the cemented down edges, doping *only* the overlap. The overlapping trailing edge is then doped down.

Fig. 8-4. When doping the overlap, use small brush and put dope *only* where overlap reaches, and will be covered by the finishing pinked tape. Do not pull fabric tight.

Notch, fold in and cement neatly around all openings, such as those for the hinges and control levers on the Fleet ailerons (Fig. 8-5). These will also be pinked-taped, of course.

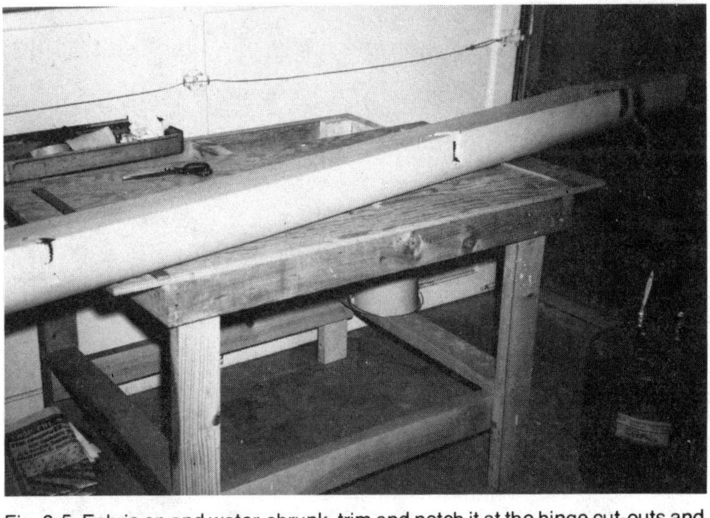

Fig. 8-5. Fabric on and water-shrunk, trim and notch it at the hinge cut-outs and cement edges inside the aluminum leading edge. It was necessary to do that at the control lever before bringing the fabric around the leading edge.

Fig. 8-6. Ailerons wet with the fungicide coat. The one in foreground is laying on a pair of thin sticks so it will not stick to the bench.

Fig. 8-7. Rear, reinforcing tape on rib cap-strips; foreground, marked and pre-punched holes for rib-stitching.

Fig. 8-8. Ribstitching is done the same way on all surfaces. A wing is easier to do hung or supported in a vertical position.

Fig. 8-9. Rib-stitching of the pair of ailerons is finished. Ends of reinforcing tape have been trimmed to meet edges of leading edge metal (front) and gussets (rear). Foreground: aileron is bottom-side up; rear: the other is top-side up, facing camera.

Fig. 8-10. Doping pinked tapes on aileron ribs. It will also go around all openings, the spanwise tapes next, and last, finishing tapes around the ends. Never pull when laying the tapes. They must be worked down close against the cover over the stitches and the length of the spanwise tapes.

Shrinking and Doping

With fabric now on the wing, or other control surface, wet it thoroughly with the mineral-free water, working symmetrically for even shrinkage. Allow it to dry completely for 24 hours.

Unless the wing can be hung for easy access to both sides, it must be turned over for each application of dope to follow, to assure even application and uniform shrinkage.

When resting flat, the wing cover will sag and lay on the edge of the rear spar until it tautens. Use the rubber suction-cup to lift it free when the dope is dry to the touch, until it tautens and remains clear. *Do not* permit it to dry, sticking to the spar. Be prepared to do this, beginning with the first fungicide coat of dope, for which the wing (or other) is now ready. Remember, this is *thin* so take care that it does not run through and dribble onto the inside of the opposite surface.

This quotation from *AC 43-13-1* is of interest and is applicable to all aircraft in the high speed category:

On aircraft with never-exceed speed in excess of 250 miles per hour, anti-tear strips are recommended under the reinforcing tape on the upper surface of the wings, and the bottom surface of that part of the wing in the slipstream. Where the anti-tear strip is used on both the top and bottom surfaces, pass it continuously up to and around the leading edges and back to the trailing edge. Where the strip is used only on the top surface, carry it up to and around the leading edge and back on the lower surface as far aft as the front beam (spar). For this purpose the slipstream should be considered as being equal to the propeller diameter plus one extra rib space on each side.

Cut anti-tear strips from the same material as used for covering and wide enough to extend beyond the reinforcing tape on each side so as to engage the lacing cord (rib-stitching). Attach the strips by applying dope to that part of the fabric to be covered by the strip, and apply dope freely over the strip.

The above would certainly be advisable for any racing plane.

Stitching

Rib stitch spacing should be as per original for re-cover of an older plane. Note that the stitches within the slipstream area (propeller diameter plus one rib either side) are only *half* the spacing of the outboard portions of the wing and other control surfaces, as is the spacing of the first two stitches aft of the front spar of every rib.

Basic and essential diagrams illustrating the acceptable methods of making the stitches and knots are given in *AC 43.13-1,*

Fig. 3.4 through *3.8*. *Figure 3.4* is the standard "modified seine knot."

With tape rule and pencil, mark exact locations for every stitch across each rib on both top and bottom surfaces. This will sound time consuming and superfluous, but it will help the work go faster and result in straighter stitches. Because of the difference in linear contour distance of the upper and lower sides of the wing, the top and bottom spacing will vary. That on the bottom can be adjusted to accommodate those on top while keeping the stitches as nearly perpendicular as possible. The top side spacing is primary.

Assuming a plywood or aluminum leading edge sheathing, start with the first anchoring stitch just aft of the front spar, for measuring.

When a person is working alone, it facilitates the actual stitching to puncture with the rib-stitch needle all of the holes on both sides, especially if the wing is laying flat. It is a lot easier to find a pre-punched hole than just a pencil mark when you are bent over upside-down, trying to poke the needle through from the top, or even if the wing is in a vertical position. Notice the holes for stitching in the aileron; the pencil marks were made to measurement but since the ribs are on the diagonal, the rib-stitches will be square across them. See Figs. 8-7 through 8-9.

The cord will be pulled over to run alongside the rib cap-strip, with the knots also at the side, tucked in as much as possible for minimum projection.

Use strips of masking tape at each end to hold the rib reinforcing tapes over the ribs, top and bottom, while doing the rib stitching. It is not desirable to dope these in place because it is preferable to be able to pull them straight after a row of stitches is completed; the stitches will hold it stretched and straight. However, it was doped to the Fleet aileron ribs for a practical reason: the cap-strips were ½" wide and the tape was only ¼" wide; trying to keep two pieces lined up side-by-side while rib-stitching would not have worked very well. Standard widths of tape available are ¼", 5/16" and ⅜". The width of the tape must be just sufficient to cover the cap-strip without extending beyond it; stitches are not to go through the tape. All stitches are to be as close as possible to the rib cap-strips.

Internal control cables should be pulled snug and tied or tapered at the root end of the wing wherever they exit (such as aileron control cables) so they will be held in their normal rigged position and will be free to operate properly after rib-stitching is completed. Otherwise, it is too easy to have them interfere with the stitching or vice versa.

To continuously stitch the full length of a rib requires a cord approximately four times chord-length. On a large wing this would be too much to handle with the repeated pulling through and tying of knots. A splice-knot, located so it will be pulled within the wing, is quite acceptable. One continuous cord will not be too difficult to work full length on a small wing of up to 50″ chord. Of course the rib depth and stitch spacing will have a lot to do with the amount of cord used; close spacing or a deep rib section will consume a lot more cord than for an average light plane wing. After a few ribs you can set up your own system.

Any time a knot "goes wrong"—and we all goof!—carefully undo it and try again, watching each consecutive move carefully to spot the error. A large chord wing cannot be worked all the way from leading to trailing edge without changing position; to work from the rear then means reversing yourself relative to the action. The knots must be made in the same way the full length of every rib.

Since the cord must be tightly held by one hand and pulled by the other, it won't be long until there are some tender spots on certain fingers. Before your own "fabric" wears thin, put two or three adhesive bandages where they will do the most good. Don't try to do the whole job at once. By the time your fingers will have toughened up the procedure will be familiar and easier and the work will go along much faster. All the knots must be precisely made, and *tight*. A poorly made knot will loosen and allow the cord to go slack both fore and aft. The purpose of the stitching is to secure the cover to the ribs uniformly for an effective airfoil to obtain the flight performance the aircraft was designed for. So the tighter and smoother the better. A good appearance is an added benefit.

Be careful to keep stitches clear of wires, cables and control mechanisms. The pre-punched holes for locating stitches also serve well as "peep-holes" to watch for internal obstructions.

Trim loose fore and aft ends of the cord to about ½″, and the reinforcing tapes where they meet the leading and trailing edge metal (or gussets); letting the latter continue over something under the fabric results in an extra and unnecessary bump. Use the triangular point of the needle to work the reinforcing tape flat and even wherever a tied knot may have curled up the edges of it, and pull gently from each end to remove any curves or wrinkles. Each end of the tape should extend about ½″ beyond the stitches and be cut off neatly with pinking shears.

Chord-wise pinked tapes are doped on the full length of the rib caps-strips. Do *not* pull it tightly but allow it to lie in contact with the cover between stitches. Dope it down snugly in hollows. Each end should terminate where it will be covered by the span-wise tapes over the front spar edges and the trailing edge. Do *not* pull the tape along the trailing edge either; any tape pulled across low spots is very difficult to work down tight onto the fabric. Pinked tape is to cover all seams, edges of any underlying parts (such as the overlap of the leading edge metal), corners at the root and aileron and flap well areas and around all fittings and openings. Areas around moveable controls and where fairings or other parts will rest against the cover should have neatly cut patches or strips of the duck or twill to protect the fabric (Fig. 8-10).

The spanwise tapes go on *last*; on top and bottom edges of the front spar and the leading and trailing edge metal. Wide 3″ or 4″ tape is used on the latter; the leading edge gets a lot of cleaning and rubbing and it is needed on the trailing edge to cover the seam or overlap and the edges of metal both sides.

If the wing is an upper panel with a navigation light mounted on the tip bow, a separate piece of pinked tape can be put on aft of the light, carry on around the bow and along the length of the trailing edge. If the wing does not have a tip light, cut the wide leading edge tape long enough to go around the bow *to* the trailing edge. One piece all the way would be too much to handle well. Dope it to the leading edge, being very careful to keep it straight. Let it set up firmly before starting around the curve of the bow. Lay a heavy, wet coat of dope all around the bow, top and bottom, get a firm grip on the end of the tape while bracing your body against the aft edge to steady the wing—then pull the tape around the bow, watching to keep it centered over the edge of the bow. The edges will lay down flat against the cover and can be quickly doped into place, resulting in a neat, smoothly contoured tip. This system works well on all rounded tips or any convex corners, unless they are of very short radius where notching would be required. It is the only way to go around them without getting wrinkles in the tapes. This same procedure will be used where the tape is restarted aft of the wing light and brought on around the trailing edge to the root; just allow more time for the dope to dry behind the light-base (for about 3″), otherwise it would pull loose when pulled back around the bow.

For aircraft with never-exceed speed of 200 mph, it is recommended that the trailing edge tapes be notched an inch deep at intervals of 18″. In case a portion of the tape should loosen it could tear free without peeling the entire trailing edge tape. If proper doping procedures are used this should not happen. It is an added precaution for fast airplanes and certainly should be done on any racing planes. It is not necessary on the average airplane.

Before continuing the doping of the wing, following the procedures described for the fuselage; mask the root ends of the spars (if you have not yet done so) and anything else to be protected from dope. Leave them covered until all doping is completed.

Give all tapes and reinforcing patches a full coat of dope to bring them up to match the fill of the base cover, two coats if needed to tauten them smooth. Sand well all pinked edges.

Refer to your wing plan for the proper locations for all inspection hole cover reinforcing rings (fore and aft at each compression member) and for access to all internal control connections. Locate them so the hole would not be cut through any of the pinked tape which can be under a segment of the ring. Brush on a generous application of dope and press them tightly down until they stick completely. Don't get dope on the outside of the rings or they will stick to your fingers instead of the fabric. Then dope over them again quite heavily.

Drain-hole grommets must be placed just ahead of any part that could trap moisture in any component of the airplane, including the aft part of any "boxed" section, the lower part of the rudder and along trailing edges of all components. They are advisable on both inboard and outboard sides of trailing edge tips of all wood ribs, especially on fairly large wings. They are essential on the outboard side of all ribs in every component; outboard, because dihedral will cause moisture to drain inboard. Be generous with grommets; they drain and they ventilate. Because of their venturi action, seaplane type grommets are far more effective than plain grommets. Remember to cut the holes for them *before* cementing them in place.

Figure 8-11 shows the home-built wing panel completed and ready for silver dope. It shows a nice gloss after six full coats. Sand very carefully, every other coat of dope, with extra sanding of pinked edges and *none over the stitches*; work carefully around and between them. Even very fine emergy paper will scuff through the tape over them. Use #180 grit emery for all sanding of clear doped surfaces. Be sure to wipe off sanding dust each time with a cloth

Fig. 8-11. Clear doped homebuilt wing panel, ready to go to hangar for silver spray. Note drain grommets just outboard of each rib; dihedral will cause any moisture to drain there. Note gloss after six full coats of clear butyrate. This is a Grade-A cotton.

lightly dampened with thinner. A minimum of full six coats of clear is needed to give a smooth gloss base for the silver and color.

The foregoing procedure is repeated for each component part. It goes quite quickly when you become familiar with the routine and develop a feeling for working with the materials.

A plywood-skinned aircraft would be done in the same way except there would be no rib-stitching. The fabric would be doped directly to the prepared plywood surfaces. Extra care to avoid wrinkles before the fabric tautened would be necessary. A full wetting with each coat will take care of this if it has been fitted well from the beginning.

Before proceeding with the silver, stop and check for any items overlooked. A little patch at this point is not too distastrous.

Covering The Center-Section

A biplane center-section is probably the most difficult single item to cover because there are so many protruding fittings, fuel tank inlets and outlets and usually a curvey trailing edge bow with cambered contour. Only the leading edge is straight. The leading edge and the aft portions are usually skinned with plywood and the bottom, under the fuel tank, may be also, in which case the fabric is doped directly to it. Or it may have former ribs to which the fabric must be secured by rib-stitching. In case of this latter arrangement, there are holes in the rib webbing to accommodate the rib-stitches, and a curved (circular) needle is used to do this. The modified seine knot is used, as in all stitched surfaces.

Note: The same procedure is followed where fuel tanks are mounted in wing panels.

So *before* covering the bottom, note the spacing and exact location of these holes for stitches. Make a diagram with accurate measurements for reference. Also note correct locations of drain holes in boxed sections for the placement of grommets. They are absolutely essential in a center-section.

Whether or not an edge-seam will be required for sufficient span would of course depend on the size, and your material width. Center-sections of Stearmans and the Waco UPFs are quite large—over 6 feet.

There is no very practical way to make an "envelope" cover. Cut a piece large enough to wrap around from trailing edge to trailing edge. Cut one end to match this contour with a 2″ allowance; this, for the bottom. Cut notches as necessary so the fabric will lay evenly and cement up around the trailing edge, starting in the middle and working outward; be sure there are no wrinkles cemented in. Then carry the cloth forward, up around the leading edge, over the top and back to the trailing edge. There will probably be one or more cut-outs for things on the way; keep them as small, snug fitting and neat as you can. As with the bottom part, cut off to a 2″ allowance, notch, and dope, overlapping the bottom fabric. This results in nearly a 4″ total overlap. For a more even finish, stagger the notches of top and bottom parts to avoid unnecessary overlap and thickness. These little things are so obvious on the finished airplane.

There should be a 2″ allowance to be cemented down, after notching to follow the rib contour as on the wing root. Then cover the ends with a full size piece, cut to fit the full rib, but cut out for the spar-ends, and wing attach fittings, etc., as necessary. Dope to the *overlap only* until after the water-shrinking.

Cut and fit neatly around the hand-holds — this is something to be done also on the lower wing panels.

Then comes the water-shrinking, and the same procedures used for the wing panels.

A center-section deserves special care. It very literally holds a biplane together. Usually it is the last thing re-covered of the wing group because it holds everything else together — if it comes off, so must all the wing panels. Unless the plane goes over on its head, it seldom suffers any damage. So it generally stays on until it has to be re-covered. Good grommets to keep it dry and ventilated are a *must*. There is not only atmospheric moisture but also possible slight fuel leakage from inside or outside the tank.

To protect the fabric at the root ends, matching the wing roots, dope on strips of duck or twill where the metal fairing rests. This fairing covers the gaps between center-section ends and wing panel roots. Doping fabric strips over these gaps is not advisable; they should be removable for inspection purposes.

Use wide 3″ or 4″ pinked tape on both the leading and trailing edges, as on the wings. It will be necessary to notch the tape to follow the inside contour of the center-section bow. Two inch tape will take care of most of the other places — and there are a lot of them, especially if the fuel tank is fitted. Try to get very close, tight fits around fuel tank filler necks so spilled fuel cannot run down inside the center-section (instead of the tank). It will also keep moisture out. The fuel tank vent, vents the tank only. The drain grommets vent the center-section.

The vital importance of the biplane's center-section *cannot* be overemphasized. All wing rigging begins there and it is totally dependent on it. A poorly rigged plane can range from just below average performance to outright dangerous. No plane can be reassembled "just like it came apart" with the expectation of its being *right*; that just never works. Each assembly must be done step-by-step, by the specifications. And rigging should be checked periodically too.

Miscellaneous Items

Many light planes have tubular landing gear legs faired with fabric covering. Arrangements vary but in general it is desirable to enclose as much "garbage" as possible for streamlining and appearance. Follow the basic procedures in putting on the fabric, keeping the weave in line with the slip-stream. It often helps to first make a paper pattern, because of angles, to fit the V shape; brown wrapping paper is fine for patterns. One piece will cover, by cementing first to the rear leg, going forward and around the front (main) leg then back and overlapping along the rear leg. There should be 2″ of the actual overlap. The top edge will also be cemented (with overlap) to the light tube member which joins the upper ends of the fore and aft legs, paralleling the bottom longerons close to the fuselage.

The bottom V at the axle is frequently done poorly. Take special care to do a neat job here. Carry the fabric as low as you can to include as much as possible of the axle cluster. It is unavoidably a very dirty area, both actually and aerodynamically. A neat covering cleans it up both ways.

It is also a place most subject to moisture accumulation, and no other part of an airplane is exposed to so much water, dirt and oil. So provide for adequate drainage. Install *at least* one seaplane grommet at the lowest possible point where moisture can drain out. Use *only* seaplane type grommets here. Water and dirt would only be blown and splashed *into* a plain flat grommet hole.

Put 3″ pinked tape all around gear leg covering fabric, as with all other covered parts; these have extra wear and they are highly visible. Doggy-looking gear fabric makes a bad impression.

Very few planes now flying have lift struts (or others) made by taping wooden streamline formers to round tubing, but some amateur-builders are going back to this technique to avoid the cost of expensive streamline tubing for lift struts and landing gears and there is nothing wrong with it.

The well varnished wood formers are taped on with pinked tape and finished in the same way all other covered components are treated. Biplane "I" interplane struts would be done in the same way. When wrapping a streamlined wood/tube strut, start at the *bottom* end and work *upward*, thus constantly closing the top edges of each layer of tape and helping to seal out dirt and moisture. Put drain grommets at *bottom-rear* of interplane "I" struts.

Use the fungicide dope on *all* covered parts, particularly these gear/strut pieces which are not well vented, being tightly enclosed.

Give them the same amount of dope and sanding as all other covered surfaces; an extra amount of both is recommended because of the actual heavy-duty handling they will always have.

Another suggested use of fabric (from the author's personal experience):

Having observed the common problem of having the finishing pinked tape come loose from the corner formed where the metal headrest meets the fuselage turtledeck, I covered the aluminum headrest of my own homebuilt with Grade-A cotton, securing it at intervals along its length with ties of reinforcing tape stitched to its lower edges. These tie across to each other underneath the headrest. The fabric was water-shrunk and doped along with the rest of the fuselage. There has been no trouble with pinked tape loosening or with paint peeling from the aluminum headrest. See Fig. 5-6.

Patching and Re-doping

See *AC 43.13-1*, Chapter 3, Section 3, *Fig. 3.10* and the *text*. This section is quite explicit in all details without further comment with one exception: once a patch of any appreciable size is secured on all its edges, *water-shrink* it to remove any wrinkles and to tauten it before starting to dope it, and *use the fungicide* in the first coat as with all fabric; a patch has no immunity to mildew.

Any patch should be doped and sanded equivalent to the original. With special care, a repaired portion can be finished so it will be barely discernable. The small patches on Fleet rudder and fin (Fig. 8-12) will not show after refinishing. The silver was sanded down to clear, then the patches were doped in place. They were repeatedly doped and sanded until absolutely smooth. The entire surface will be resanded before being resprayed with silver. They will then be sprayed with color (black) to match the rest of the fuselage.

Patches are no problem with butyrate finish, but enamel is just *trouble*. Butyrate must be used to build up the fill of the fabric

Fig. 8-12. Small patches on fin (from a slipped tool) and rudder (from stones at a rough field). Linen patches on linen, doped and sanded until completely smooth. Surfaces will be sanded over all before re-doping. For an interesting sequel, see Fig. 10-1.

before finishing the patch with matching enamel, and butyrate *lifts* enamel where they meet around the borders of the patched area. It goes on and on in an ever-widening area until you feel like the "Sorcerer's Apprentice," trying to stop it. Just try to keep the dope very light where it overlaps the edge of the enamel, and sand the area well before re-spraying the enamel.

The finish on a synthetic must be removed for a large patch in an area big enough to use an iron to shrink the patch, i.e., beyond the cemented over-lapped edges of the patch. If the hot iron touches the finish (dope, lacquer, enamel, etc.) it would stick or wrinkle and be a real mess.

Regardless of how careful we try to be, there are the inevitable little holes and scuffs to be patched from time to time. Rematching the color is one of the attending problems with patching as all colors do fade to a certain extent.

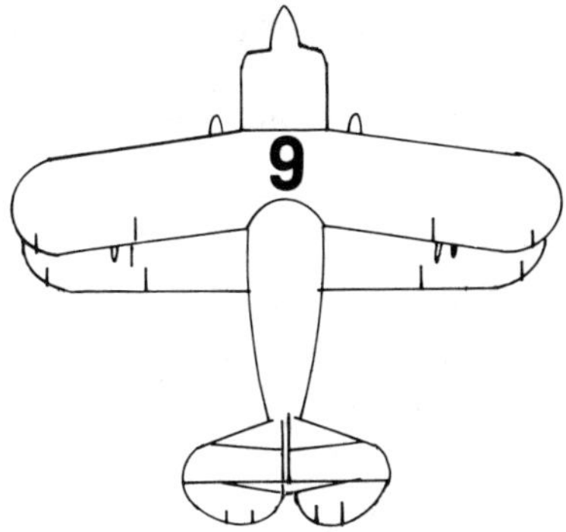

Spraying Silver Dope

Silver is a magic word. It signifies that the end is in sight; these parts do look like they belong to a real airplane that will fly pretty soon. During the long, long hours of tedious work, all thought was concentrated on nothing but the work. Ever getting to *fly* the thing seemed so remote that dreaming about it was a luxury you didn't dare indulge in. Dreaming just didn't get the job done. Anything but working on the airplane was a waste of time, or so it seemed. So stay with it a little longer now; be patient; be careful.

Silver is used to cover *all* fabric on the airplane, regardless of the final finish color, because it provides excellent protection from the rays of the sun which cause fabric deterioration. Very often the silver itself is used as the finish on all or part of the airplane.

The old gauge of how much silver is enough was to shine a bright light on the outside; if it showed through, more silver was needed. This is still true, but we are not concerned with bare minimums—we want a good, long lasting finish. You may have read of two coats being minimum. Maybe so, for someone trying to "just get by;" but that is *too minimum*. We prefer at least double that amount, with four full cross-coats when it is to be an under-coat for color and *six* if silver is also the finish.

The clear dope forms the base. Silver provides the protection. The color is primarily for appearance and it also vastly increases the visibility of the airplane, which is more and more important with the increasing numbers of aircraft in the sky.

The price of a beautiful lustrous finish is a slight weight penalty, but a very nice finish can be achieved with a reasonable amount of dope properly applied each step of the way. A too-heavy accumulation of dope can be so thick that it will harden, crack and peel, thus defeating the whole purpose. A moderate amount will remain smooth, flexible and alive for many years with proper care. "Just enough to get by" is a waste of time, effort and materials. You can judge the integrity of a shop or mechanic by the finish put on an airplane. The part that hurts these days is that so few licensed mechanics know a good or bad job when they see one. To check dope and fabric on a plane you or a friend may consider buying, use the old test of shining a light on the surface while looking through from the inside (this takes two people). If it shines through, someone is trying to sell a cheap dope job. On the other hand, a hard, glassy, brittle surface is even more objectionable; it will crack and peel. If the fabric is still new (check A/C Logs), you *can* sand and re-dope but there really isn't much that can be done where there is already too much. So with your own work, be conservative; it will give the best overall results. You want it to look like an airplane, not a car. A shine can be put on later with a good polish, if you want that.

Back to work now.

Unless there is a separate shop, hangar or other suitable place to spray in, one must be made. The pliofilm partitions in the basement of Mr. Hall's home (in many of these photos) worked perfectly; heat and de-humidifier were available and there was an exhaust fan in the upper left of one window. Work was continued in spite of high humidity (rain) outdoors, during all of the clear-doping period. Sprayed dope does drift, but it dries before it goes far, and settles as dust. (Sprayed enamel settles wet and thus presents a problem.)

Unless ambient temperature is up to 72°F., there must be heat—but *no flame* because of fire hazard. There must be some kind of ventilation to exhaust the pigmented dope which would otherwise hang in a fog. But again, avoid drafts to preclude blushing. Because spraying atomizes the dope, it blushes much more readily than when it is brushed.

The air compressor must be capable of maintaining 50 to 60 lb. pressure at the nozzle; enamel requires less than half that, but dope is of heavier viscosity and dries much faster; it goes on in greater quantity.

There must be a pressure-regulator in the line incorporating a water drain valve at its lowest point. Do not let the regulator lay on the floor because the water trap would be ineffectual. Water droplets in the sprayed dope result in bubbly blemishes which break down into little craters which spoil the finish. Open and drain this valve-outlet *often* to prevent accumulation before it gets into the hose-line to the nozzle.

A small pressure or suction type "cup" is adequate for spraying of trim, stripes, numbers, etc., but it does not put out well enough to spray an area of any size. The small capacity will not permit sufficient continuous flow or volume.

The best type of equipment is a pressure-pot used with a spray gun (nozzle). The nozzle is adjustable for both amount and mixture (air and dope) delivered, and for the "fan" pattern, to a much greater degree than is the cup-nozzle. There are two common sizes of pressure-pots: 2 gal. and 5 gal. The 2 gal. pot is fine for anyone not doing a very large amount of work, where a 5 gal. size would be a convenience. There are commonly available through commercial equipment rentals. Good quality paint spray equipment is quite expensive to purchase. The DeVilbiss Spray Equipment Co. publishes several excellent booklets on selection and usage which are well worth reading. If you rent, try to be sure of reasonably good weather for the period and plan the work so it can be carried along continuously. (Optimist!)

Make sure the entire system—pot, lines and nozzle—are *clean*. Run thinner through it, let it soak if necessary, and keep running the thinner through until it comes out consistently clear. (Retrieve the thinner in a 5 gal. can; even if not clear, it is good for cleaning purposes.) Any residual contamination would spoil the work.

A wide-mouth 5 gal. can is preferred for mixing the silver dope. Mix ½ lb. of silver paste with a gallon of butyrate thinner. It takes a lot of stirring to get it completely dissolved and mixed. Add three gallons of dope and continue stirring until there is no thick pigment on the bottom. This amount will allow space for more thinner if it is needed, or if it seems too thin, more dope can be added. The proportions are rather indefinite due to differences in pressures, nozzles, weather conditions, etc. I have found it expedient to vary the mixture from one day to the next to get good flow and spread. A full oval "fan" of sufficient body to give a good build-up is the objective. The spray gun will become clogged and the dope will come out in "bloops" if it is too thick. There will be similar problems if the pressure is too low. If pressure is too high, the dope will be dissipated in a too fine, dry mist. It should be obviously wet as it goes on the surface.

Pour the mixed silver (and all pigmented dopes) through a paint strainer into the pressure-pot. Any tiny lumps or foreign matter will foul the small orifices of the nozzle (so will thickened dope), and this requires complete disassembly and cleaning of the gun before work can continue. If you do not have a helper to hold the paint strainer while you pour, rig some kind of holder for it. It runs through pretty slowly; to try to hold a 5 gal. can of dope with one hand and a paper strainer with the other, while pouring slowly without spilling it, is a bit of a challenge. Have plenty of newspapers under all the dope cans and equipment; there *will* be spills. Whenever it is necessary to leave the work for an extended time (more than an hour or so), run thinner through lines and nozzle and leave it in there until work is resumed. This will keep the equipment clear of dry or thickened dope and will save a lot of time.

Before starting the silver, give the clear doped surface one more light, overall sanding, and as always, wipe off the sanding dust with a slightly dampened (with thinner) rag. Check again to see that everything is masked off that should not be sprayed with the silver.

A "spray coat" more explicitly means a cross-coat which in a way is a double coat. With the nozzle adjusted for a good full pattern, it is passed back and forth over the piece for a complete covering in one direction, and then back and forth again at 90° to the first set of passes. For example, let us say you are spraying one side of a wing (either hanging or resting flat). One full cross-coat would consist of spraying it completely with chord-wise strikes, then again with span-wise strokes, or vice versa.

The nozzle should be held 8″ to 10″ above the work. Begin by shooting into corners and along edges before going over the main surface, with *every* coat, to insure these places will not be neglected and for more even coverage. Carry the spray gun along smoothly, aimed as directly as possible at all times, at a constant elevation above the surface, to lay a smooth, even and *wet* swath of pigment. Do not expect a full cover all at once; allow it to build. Each pass of the spray gun must somewhat overlap the previous one for an even cover. Try to vary the overlaps' locations to avoid "striping". The purpose of the cross-coating is to achieve an even coverage. An extra pass with each coat along all leading edges will provide more protection to these exposed places.

Trigger the spray on slightly *before* it is quite over the piece, then off *after* the spray has cleared the other edge of it; in other words, the gun is triggered both on and off *clear* of the work. In this

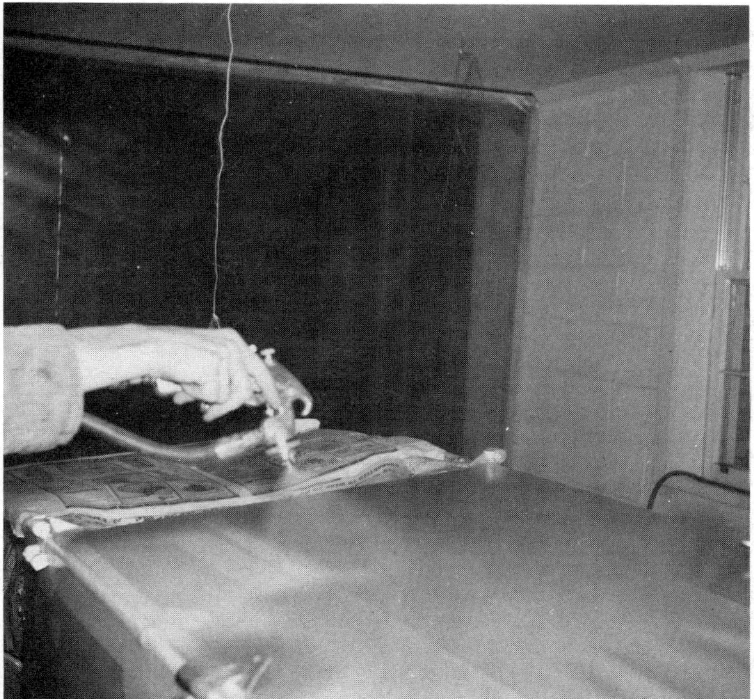

Fig. 9-1. Spraying silver butyrate, keeping the nozzle at a uniform elevation and attitude above the work. Slight wrinkles visible in lower right corner indicate very good penetration of the dope from the full spray; it will retighten immediately. The fan pattern is set for a vertical position (photo by Ben I. Hall).

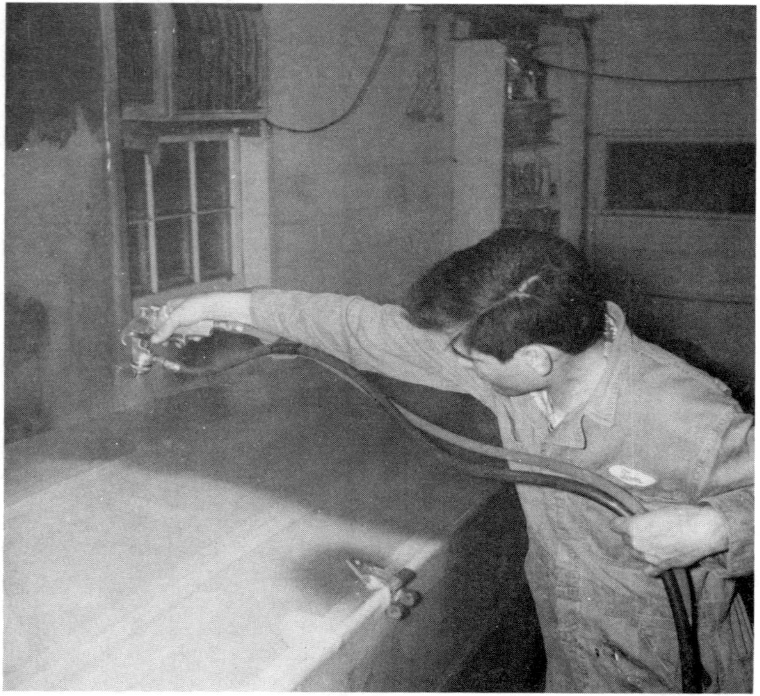

Fig. 9-2. The area close in around the fittings has been sprayed first, as are the edges. Hose and clothing must be kept clear of the work. Exhaust fan is in upper part of the window. Wetness of the sprayed surface is visible under the gun and hand (photo by Ben I. Hall).

way, coverage will be constant and uniform, and if the nozzle should drip when it is shut off, dribbles will fall clear of the work. Hold the spray gun at the same level and attitude all the way; do not swing it in an arc as you spray (Fig. 9-1).

Spraying is tiring. It helps to have the work resting lower than you would for other activities, like repairs, because the spray gun must be held at least 8″ above it and it is necessary to lean across it for good coverage. Of course, a large wing must be worked from both front and rear, if laying flat. Observe the illustrations of spraying silver on the fuselage.

Hold the hose clear of the work (Fig. 9-2); if it drags across the wet dope it will spoil the whole thing. Be certain also that no part of your clothing brushes over the doped surface. Speaking of clothing: wear old "run-out" pants and shoes. Silver and/or colored butyrate will go through hundreds of washings and ironings and out-last the pants. Thinner will remove dope from shoes but it also takes the finish off the leather. A respirator is recommended, especially where there is not good ventilation.

To avoid the effects of dry overspray ahead of the spray pattern, keep working "forward", not back away from the wetted area, so over-spray ahead of the nozzle is wetted as the direct, wet fan of dope passes over it. A dry overspray onto a still-wet surface results in a very rough, sandy texture. The relatively wet overspray on any dry, doped surface has the same effect. There will be a similar result when the nozzle is held too far away or at too much of a slant to the surface. It must be removed by thorough fine-sanding overall and then re-spraying, directing the fan-pattern more squarely on the area.

Never do any spraying of dope or paint where dust and dirt is blown about. One common annoyance are flies and bugs that make their "last landings" on fresh-doped surfaces! You can only lift them off as carefully as possible with the least disturbance of the soft dope. If it resulted in a pretty bad spot, let it dry enough to sand lightly, then hit it again with spray. One would think that bugs would be repulsed by butyrate dope but they seem to like it. *Do not use oil-base bug sprays in the work area.* The fine droplets would settle on the doped surface and that would be more objectionable than the bugs.

The silvered surface will show up little—and not so little—blemishes of which you had not previously been aware. If any short-cuts were taken in earlier procedures, this is where they will become apparent. And to some extent these will determine just how much silver will be needed, over and above the normal amount. If the surface is rough for any reason, sand and re-spray, a couple times if necessary. Any defect showing in the silver will show *at least* as much in the color to follow, though white seems most forgiving. The *first* coat of silver should be sanded almost away (#180 grit emery), for a start, then after every other coat, until final, described in detail in the last chapter.

The following word description of the adventure of spraying silver dope was written by a fellow lover of antique-type airplanes, writer by profession and flyer by passion, Dick Bach. It first appeared in the January 1970 issue of *Air Progress* magazine.*

"These words are written under the influence of dope. The basement is a glue-sniffer's paradise; the room is a rolling silver fog of cellulose acetate butyrate dope, and in the midst of that fog, on saw-horses, rests a pair of Aeronca Champion wings. Hanging from the ceiling are the tail surfaces. All silver.

"It is strange stuff, dope. You lay it on for awhile, cross-coating the fabric under the spraygun, and you don't notice that the air is filling with millions of droplets of dope-vapor. Breathing it, you begin to think strange thoughts.

"There's the spray pattern glistening wet, moving back and forth, pelting down in a little sea over the weave of the wing fabric, and reflected in it are fragments of ideas and flickers of mind-images darting like butterflies, more of them the longer you breathe.

"Numb, I'm getting numb. I can't feel my lower lip when I bite it. But that silver sea goes on the wing, by Ned, that sea goes on. Each spray stroke takes four seconds, and then another one overlapping for four seconds more, coming back. The sea widens.

"I'm losing my sense of balance. I'm teetering down here like some drunk, doggedly spraying these wings. There is so much work to do to get one little airplane finished and in the air. So *much* work. Does the world know how many moments of thought and act and effort there is in just one light airplane? Nobody has the dimmest idea, except the man who loves one.

"Spray—spray—like the sound out of yesterday. Is it worth it, Richard? You've neglected all the things you should have done, for this airplane. You've put off finishing the story in the typewriter upstairs, you've got letters waiting a month that haven't been answered. The lawn outside is four feet high. Bills are coming in.

---

*Richard Bach, *Reflections In A Silver Sea* (New York: The Macmillan Co., 1969).

You know that you must write, if you wish to go on eating, in this household. But what are you doing? You are down here in the basement spraying dope in wide patterns, your arm a slow pendulum, back - and - forth.

"I feel lightheaded as one of Stuart MacPherson's homemade hot-air balloons. Remember the time Stu sent us that balloon in the mail, and we unfolded it and it was six feet tall, light as cottonfluff? Rick Drury was here, and we took it out in the snow and built a big fire out of newspapers and we flew the balloon. Somehow it got named the *Joyful Expression*, and it went up a hundred feet like the tissue-paper spirit of the brothers Montgolfier, and came down at last, drifting into a balloon-eating tree. Took an hour to rescue it.

"Spray—spray—like the sound of a propeller turning—quiet after the engine has been switched off. Of course, it's worth it. On the way out to the airport, I looked out at the fields and they were all green and level, and I thought how fine it will be when the wings are done, when this little airplane is finished at last after these months of work and Bette will have her Champ to fly again. We'll fly together, the biplane and the Aeronca, the way we did after she soloed, before we took her airplane apart for rebuilding.

"Spray—spray—the sound of wheels flicking through grass a second till touchdown. So I'm escaping, is that it, with all this dope that makes the room tilt and reel about me? Sure I'm escaping. Sure I'm getting away from work and bills and chores. But there's a magnet out there drawing me, and I say I'm guilty and I call this escape only because I can't see farther than the end of a spraygun. The only thing that matters in life is flying, Richard, isn't that true? Lifting myself up over all those odds and bills and chores, all those balloon-eating trees that would drag me down. I hereby refuse to let myself be dragged.

"Spray—spray—the sound of hours and days and months going over the falls called NOW. There's the obsession that drives me—to get this airplane back in the sky so that we don't have to be staked to the dirt.

"There. One final pattern of the spray and another coat is finished, shining wet silver, and I stagger upstairs. Even now my typewriter spins slowly under my hands, and it's odd to watch the designs in ink on paper.

"Patterns in ink, patterns in paint, patterns in the way we choose to live our lives. To see these patterns, to watch them unfold in a dope-fog basement while an airplane comes back alive, I think there's more to flying than can be learned in a long, long time."

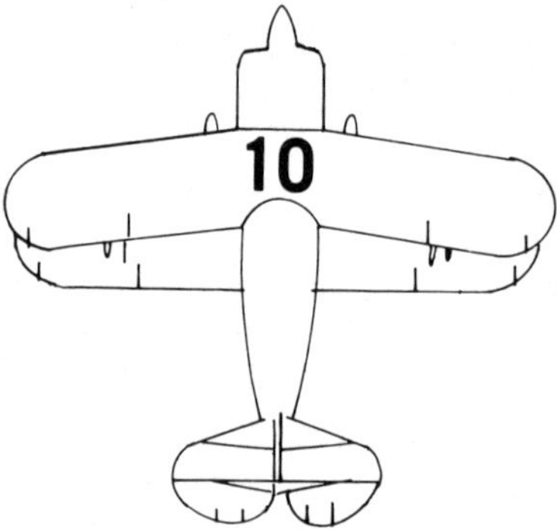

10

Spraying the Color

Silver gives the best protection to the fabric from sunlight and for that reason provides the longest lasting finish, but unless a highly visible color is used with it, it is the best of in-flight camouflage, especially in a hazy atmosphere. The greatest danger at the present time is being overtaken from the rear—our "dope and fabric" planes are generally slower than new production aircraft—so the view of the rear needs to show eye-catching color. The front is not so critical; you can "see and avoid." *Earth* colors are useless; the plane may look nice parked on the ramp, but their visibility on the ground (in case of forced landing) and in the air with the ground as a back-drop, is simply nil. With the ever-increasing air traffic, this consideration is very important.

To seal the silver, as finish or for undercoat, following the last full coat, sand completely with #400 grit emery, very lightly, then mix the final sealing coat of clear (remains of silver in the pot won't hurt at all) with about 20% thinner. A residue of silver in mixing can or in the pot would *not* be acceptable over another color but it will not hurt at all over the silver. Spray this clear mix over all, *wet*. This can be done following any color you use but it is particularly effective with silver. This final, clear coating will result in a deep, glossy finish that will look "new" for years. This simple procedure is common knowledge to many model airplane builders and it works equally well on the big ones. If you do this over any color other than silver, be sure the silver is completely out of the spray system; use *absolutely clear* dope.

During the periods of time between sessions of doping, be sure to run straight thinner through the hose and gun to clean them, and leave thinner in them if you plan to resume work within a few days. For a longer time, clean and dry the system, including the separate parts of the nozzle. This way, it will be ready to use without wasting most of the working period cleaning equipment. *Never* allow dope or paint to dry and harden in the pressure-pot, lines or gun.

As with many tasks, the actual job is easy; it is the preparation that is slow and difficult.

Color Schemes

By now, if this is your own airplane, many dreams and much thought about how it is to look have been with you while you worked. If it's a professional job for someone else, you want it to *look* professional; a plane is right there for all the world to see and to judge. The paint design and the colors used are the owner's choice.

The workmanship may be flawless, but a poorly selected color combination or clumsy design can spoil the whole effect. Conversely, we have all seen an attractive paint job—from a distance—that proves very disappointing with closer inspection.

There is one certainty: a person never goes far wrong with a simple paint scheme using pleasantly contrasting colors—and not too many of them. Don't misunderstand; there is nothing wrong with being original. Any fabric covered airplane deserves to have an individual color and design. Modern production plane designs on dope and fabric planes are seldom appropriate and they do not make an old-style plane look "new"—who would want it to? We are proud of the old planes. A "jazzy" pattern on a small airplane detracts from its pleasing lines and contours and spoils the total appearance. The eye runs back and forth over the nervous paint scheme and the airplane is "lost."

It helps to make trial and error color sketches on scale three-views. Have them where they can be studied at leisure. Very often, just a slight change of angle or contour of a trim line will make a great change and improvement. Don't forget to include the bottom of the plane; that is the part seen by everyone but you when you fly. Study color samples. Look at other airplanes and study them to determine just what it is that you do or do not like about their appearance, especially those similiar to your own type. This is often a very elusive thing. Areas of darker color look more massive and heavy. Bright colors are more "aggressive" and tend to dominate. A band around a long fuselage can interrupt the length and thereby tend to reduce the appearance of great length.

The use of the plane, whether it is to be hangared or tied down outside, whether you plan to keep it for long years or to sell it, even the kind of terrain over which most of your flying will be done, are all worthwhile considerations in deciding on the color and design. We assume that all parts of this refinished and re-covered airplane have received the best workmanship, that you want to be proud of it and to enjoy flying it. That is what it is for. This is as it should be.

Once again, back to work. This time, the masking of the component parts, one at a time. (Control surfaces are always removed for repainting an airplane.) Masking tape should not be left long on fresh silver dope before spraying. If there is to be only trim color over silver, you get off easy: mask everything *except* where the trim color will go.

If there is to be only limited trim over another base-color, spray the base color over all. Allow it to dry, then mask everything except where the trim color will go. One exception: if white is used with red, do *not* spray red where the white is to go; reds have the unpleasant characteristic of "bleeding through" white. Reds can be used over white but *never* white over reds. If white trim is to be used with a red base, mask and spray separately.

If there is to be a two color combination of approximately equal areas on any component, mask one and then spray, then mask the other and spray, consecutively. Allow sufficient drying time between colors before masking over the fresh dope so the masking tape will not lift it; one day between colors is enough time.

For any additional trim such as the numbers or pin-striping in a third color, all previously sprayed parts must be well masked.

N Numbers

For a period of years, through the late 1950's to mid-1960's, there was much confusion, agitation and animosity, when desk-pilot bureaucrats decided that all civil aircraft must have 12" high "N" numbers displayed on the sides of the fuselage. The "antique" category has been more fortunate, although some owners do not yet seem to be aware of their (legal) exemption, nor do many IA's know of it; so we see a number of newly restored old planes carrying the large fuselage numbers.

The most recent FAA definition of "antique," as applicable to this question, is for an aircraft to be thirty years old. The Antique Airplane Association considers a pre-WW II airplane type an "antique," or "classic."

Adhesive-backed masking stencils are available for the 12″ numbers and they are far superior to attempting to mask them by hand. They are easy to use and work very well. Slanted and "decorative" numbers are not legal *nor* are they particularly attractive on an antique aircraft. Be sure numbers are laid out straight with line of thrust; use the top longeron or other straight member as a reference for lining them up properly.

They are to be a contrasting color. With good planning, even the fuselage numbers can often be used to enhance the paint scheme and total appearance, particularly on a relatively long fuselage. But we would much rather see them back on the tail where they belong.

Plan the entire sequence in advance so *all* the areas of each color used will be done at the same time with the same batch of dope, for good color matching.

Newspapers are all right for masking, but use double layers; wet dope will go through only one sheet. Be sure to put masking tape over the little holes along the edge of newsprint made from rolling through the presses. Mask any little tears and any gaps where dope could blow through or under. Press down all edges of tape securely along the paint line. Do not mask very fresh, soft dope; when removed it could lift off the base color and even the silver. Allow at least overnight for one color to set up before masking for the next. Do not leave tape on freshly doped surfaces any longer than necessary, for the same reason as above.

When removing masking tape, pull *away* from the line of freshly applied dope, so the tape won't peel it off, after giving it a little time to "set."

For masking close-radius curves use a narrow tape. The ¾″ standard width is best for most work. Be sure there are no openings or gaps through which spray, or overspray, could drift. Dry overspray which settles some distance from the work can be dusted off with no trouble, but wet overspray, adjacent to the pattern of the fan results in a surface like sandpaper, so as said previously, spray *forward* across the surface to keep a wet fan pattern, and do corners and edges first.

With the end now clearly in sight, be careful; don't foul up the job. This is the icing on the cake. It is what you will be looking at for a long time to come and what your workmanship will be mostly judged by. And it is gratifying to step out at some strange airport and receive sincere compliments on a very attractive airplane, especially when you did it yourself. It makes it all worthwhile. All those long tedious hours of fitting, pinning, cutting, stitching, cementing, more stitching, doping, sanding, rib-stitching, taping, more doping, and sanding and doping and sanding. Sore fingers; numb brain; tired. Piece after piece. Again and again. Relax now and then, and try to forget the whole darned thing. But the dream won't let you rest long. You want to *fly* that airplane.

Blushing

This is a good time to show you what can happen when you do *not* wait for good doping weather. It does not pay to be impatient. Take a look at Fig. 10-1, showing the rudder and (at rear) fin for the Fleet. You will see that fin is black, as intended. The rudder is *supposed* to be black too. This is an extreme example of *blush*. The answer is to wait until tomorrow and shoot it again with sufficient dope to penetrate well so it can dry properly clear through. Watch out for drafts. Don't try to spray dope in high humidity, or in less than 72°F temperature.

So it is ready now. You gather your most trusted buddies, go to the airport, and very carefully now assemble this wonderful, new-old—or really brand-new—airplane. And there it stands. (Fig. 10-2).

Your A&P mechanic and the IA, (or if it is Amateur-Built, your FAA Inspector), will come for a final inspection. "Looks good," they will say (if it really does). Don't expect extravagant compliments here; they know how it should be and they *expect* it to look good. But if you *have* done a good job they generally take pleasure in saying so, briefly, anyway. They appreciate the work involved. A sloppy job tells them more than they care to know!

Beyond the satisfaction of a neat, smooth, good looking finish, it is a lot easier to take care of, which is simple to understand.

So in every way, you will be rewarded for a dope and fabric job well done. You might even do another one some day. Yes, it is a lot of work; so is refinishing a boat or a house or anything else done by hand. With airplanes, there is the added discipline of accommodating established Airworthiness Standards. Always consider this factor from the perspective of *self-interest*: these standards are for your own well-being and safety. Nobody made these up for fun and games; they were developed through hard experience. These standards pertaining to the basic wood structures and to covering with dope and fabric have changed very little over many long years of proven practical use. Welded steel frames have been "standard" since before many of us were born. The new materials now on the

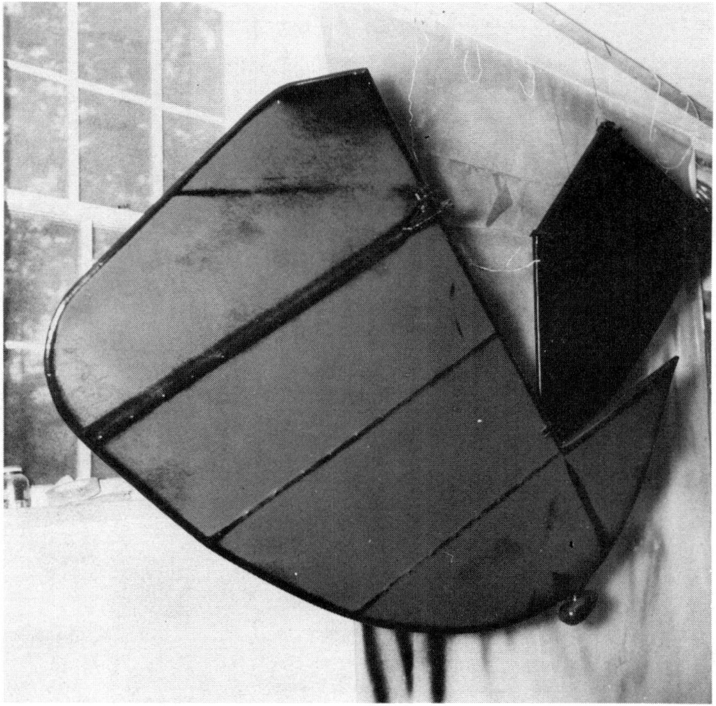

Fig. 10-1. This shows an extreme example of "blushing." The color being used is black (see fin hanging in background). It will stay this way until it is sprayed again under better drying conditions, when a good wet coat will penetrate, dry evenly and have true color (photo by Tom C. Street).

Fig. 10-2. Newly assembled homebuilt covered with Grade-A cotton; all-wood. Silver and International Orange with black details. Ready for engine cowl.

market do not alter the fundamentals of covering an airplane. Whether or not other materials may sometime make dope and fabric obsolete is a question for the future. It is not at this time a fact. A generation ago, "wood airplanes are gone forever"—yet today, in 1981, there are more all-wood and part-wood airplanes being built and flown than at any time since the First World War.

Long Live Dope!

Many of us still prefer dope and fabric covering to any other kind and it would be unwise to make any preductions for its demise. There are those who criticize the continued use of wood and of dope and fabric because it is "antiquated" and "not progressive" and "it is just doing the same things over." Some things have been *done over* since time began—does that mean they are bad? We have observed that the individuals who say these things are relative newcomers to aviation. "Antiquated" or not, just what is wrong with doing something in a way that may be old-fashioned, but certainly is well-proven and from which many people derive sincere satisfaction? By what authority does one man have the right to condemn the choices of others? Must everyone whistle through the substratosphere from A to B under the control of a little black box?

Some of us like to fly slow enough to look at the earth and contemplate the ways of man and the patterns of life which are so different in the various geographical sections. Perhaps in the era of computers it is even old-fashioned to *think?*

One's choice of a personal airplane has to do with his own needs and desires, and above all, his sense of values. There is no reason to assume that the man with a tired old Cub does not appreciate new developments and equipment and the genius of his fellow men; he just enjoys flying his Cub. It is his friend. It takes him where he wants to go; in it, he takes pleasure in the going. For all you know, he may use a business jet in his everyday work.

Nobody in his right mind has ever claimed much comfort or convenience for an open-cockpit airplane, but it is equally true that anyone who has never flown in one has never flown! That is flying in its best and most exciting form, without any special maneuvers; just flying. In the open airplane there is no false insulation or isolation from the elements. It can be startling to someone not accustomed to it. There is just the machine and the air and the earth and the flyer. And in that rushing air, you know you are flying.

As I turned off the turf runway of our local airport in my little plane one recent evening, a friend followed me in, floating softly and gently to touch-down in his pretty 1940 Taylorcraft. He and his buddy and their families enjoy this nice airplane. The two men are now completing a high performance sport biplane but they will keep their T-Craft. They enjoy it. Both of their planes are covered with a synthetic material and both are very well done.

Personalities of people designing, building, maintaining and flying airplanes are as various as the machines themselves. It would be a sad day when this would not be true in these United States of America, as it is true in some countries. Let us do all we can to keep this freedom to fly, each in his own way.

Modern production aircraft are often more practical and they certainly are a necessity for business use and for mass transportation. But when we want to fly for the sport and sheer pleasure of flying, just nothing compares to one's own airplane of one's own choice. To have built it or re-built it yourself adds a dimension of personal accomplishment that is beyond description to someone who has never done it. Yes, it is worth the effort.

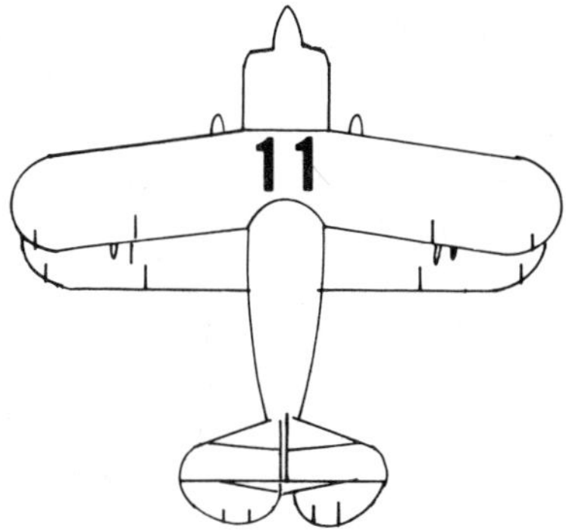

11

Changes

In this unstable year of 1980, ten years after the first edition of *Aircraft Dope & Fabric* was published, it is necessary to reevaluate the subject from the standpoint of costs and availability of materials. It has developed into a compromise and making the best of what is now available rather than choosing what I consider is actually the best. We have in no way changed our opinion of the quality of linen or Grade-A cotton compared with any of the synthetics and we will use only the natural fiber materials on our own airplanes. Also during this ten year period it has become apparent to many others that the synthetics leave a lot to be desired. The primary problem with Dacron remains getting the finish to stay on and cracking of the finish. Some ultra flashy little numbers that were trophy winners when they were brand new have not been seen on the show circuit recently.

The troubles in Ireland, along with other economic reasons, have almost priced linen off the market. It takes a friend with connections to get it at all. It is still the best material in my opinion, considering strength, finish and durability and once again I will use the 1940 Stearman, owned by W. E. Spencer for thirty years, as an example. The entire wing group and the tail surfaces have been fully restored and recovered since 1964. A couple of years ago they were all rejuvenated: sanded down, given a full cross-coat of rejuvenator and re-sprayed with silver butyrate dope. In the meantime, that unrestored fuselage still wears its 1958 Irish linen, faded blue and silver cover. It still tests for strength. More important things come first, like that oil-leaking 300 hp Lycoming!

Rejuvenating

This is an appropriate place to discuss "rejuvenating." First in importance, it must be clearly understood that *nothing* rejuvenates tired old fabric or any type of covering material. If that has deteriorated to below minimum tensile strength, it *must* be replaced. But if the cover tests well within limits while the *finish* is poor, the finish can be rejuvenated *if* that finish is dope. No type of enamel can be rejuvenated. Rejuvenator is made to be used with dope; it penetrates and revitalizes the dope.

Sunlight and general exposure tends to dry out the doped surfaces—and enameled ones too—over the years, even when a plane is hangared, but much faster if it has to sit out in the weather. I want to mention here that Spencer's Stearman *has* spent at least fifteen years outside with that fuselage fabric. He never loses an opportunity to point this out to people who say linen won't hold up as well as synthetics. Anyone can easily compare the texture and finish of the upper (exposed) and lower (sheltered) surfaces of a wing, for example. The undersides look new and retain the soft, smooth and pliable feel compared with the dry top side.

To rejuvenate any doped surface, take the following five steps:

(*Note*: Rejuvenator is made for butyrate dope and for nitrate dope. Use the correct type; they are not interchangeable.)

1. Wash the component absolutely clean, using a degreasing compound, then a mild soap and water wash followed by a final rinse and wipe-down. There must be no residue of wax, polish or oil.
2. Thoroughly sand the hardened top surface layer which will probably be dry and flaky. This is an excellent opportunity to sand out any rough spots, apply any needed little patches, feather out protruding pinked tape edges, etc. In other words, prepare the surface for a super smooth finish that will look even better than it did originally.
3. Using the dope spray gun, give the surface a cross-coat of the rejuvenator, full and uniformly, just as it comes from the can. No dilutant is used. To avoid any misunderstanding, *both sides* of every component are given the full treatment: sanding, spraying with the rejuvenator and all. Do not neglect edges, hinge wells, etc.
4. If the cover sags after the applied rejuvenator has dried, apply a full cross-coat of clear dope to renew the tautness. This is not needed if there is no appreciable sag, although it

does make a nicer surface for the fresh pigmented (color) dope.

5. Finish with a full cross-coat of pigmented dope of the basic color. Of course, if there is more than one color involved, mask off and spray each separately over the rejuvenated surface as per the original color underneath.

If you wish to completely change the color scheme, step 2, the sanding, must be done more extensively, preferably down to the original silver under the old color, then proceed with the above steps. You will have a completely new paint job as described in Chapter 10. If this is being done over Dacron or other synthetic, be certain that you do not get any more build-up of dope than was on the original cover. Better to omit the cross-coat of clear dope preceding and following the pigmented new color coat to avoid the excessive build-up on the surface which would soon result in the cracking and peeling of your new finish. To last, that finish on synthetic surfaces has to be kept thin and pliable.

If the completed finish is light weight (in thickness), one coat of clear dope, sprayed over the finish color will produce a nice lustre and will seal and protect the finish.

Remember: Reds bleed through whites and yellows, even very tiny amounts of red. It must *always* be completely sanded out and washed off with thinner, buty or nitrate depending on the dope being used, before over-spraying with either white or yellow.

This ability to rejuvenate dope finishes is an important consideration when making a decision regarding cover materials.

To talk now about the use of nitrate dope seems like a contradiction. No, we would never use it on cotton or linen for the reasons given earlier in this book. But it has to be used for the early coats applied to the synthetic Ceconite coverings because it has proven more adhesive with this material for over-laps, tapes, etc. The following fill coats, including the silver and pigmented finish, are butyrate, which is less flammable.

The removal of old finish from an entire airplane really takes more time than it is worth. We can not advise the use of paint strippers on any fabric. You don't know what effect it may have on the fabric. Paint stripper on metal parts is a great help but it could be very destructive to any cloth and not worth the risk. If an old finish peels off readily in sheets, there can be only one reason: it was not put on properly. Peeling it off might be one way to go if it is that bad. In this case, remove it *all*, and start all over just as if it were newly applied cover, using the manufacturer's approved method, applicable to the fabric. The Aircraft Log and/or Form 337 should state precisely what the plane had been covered with.

Enamels

There just is no simple way to refinish any of the enamel-type finishes. They are expensive to start with and nothing can be done to "fix" them if they start cracking or chipping. They are not compatible with dope thinners or rejuvenators. So when and if they go, the only answer is a recover job. This applies to any enamel or lacquer finish. These things were developed for automotive use and that is where they belong—on metal, and there is a lot of that on any airplane: struts, gear parts, cowling, etc.

Enamel must be completely removed in the vicinity of a fabric patch, as described earlier in this book. The patch will be put on with the kind of dope applicable: with butyrate if the fabric is natural fiber (cotton or linen); with nitrate if synthetic, or with Stits' materials if the cover used the Stits Process. *Do not mix one type with another.* The fill is brought to the same state as the original, sanded as necessary between coats, and finished to match the surface it is on. Don't expect the new color to match weathered finish; you can only approximate it unless it is all quite new. Where the new finish has to blend in with the original, with the probability of dope raising the surrounding enamel, spray very *lightly* with the pigmented dope (or clear dope while in the build-up stage) around the patched area, over-laying the enamel edge in stages. If it goes on very lightly it can dry quickly without starting to dissolve and lift the enamel. Build up the color over the patch in this way. Otherwise, the surrounding enamel would be progressively lifted and removed with no place to stop.

The only place for any enamel finish on a fabric covered airplane is strictly as a last resort: to keep the machine flyable and presentable until the end of the current flying season. Then strip it down for a full recover job. This is in case the fabric barely tests and does not justify rejuvenating a dope finish. If put on over synthetic fabric and *if the finish is not too thick,* a very light (in thickness) spray coat of enamel may hang in there for a couple more seasons. But where the finish has started to crack or chip over Dacron, it will only go faster with more enamel added.

Caveat Emptor

Those of us who prefer cotton or linen for our airplanes are being trapped between two evils: over-selling of the synthetics with the ubiquitous propaganda about their very questionable "superior" characteristics on the one hand and the concurrent unavailability of aircraft quality of Grade-A cotton *and* Irish linen on the other. It is not hard to understand merchants' desire to sell higher priced products and to drop those having a lower mark-up value. We can also see why a product selling in small quantities may well bring a higher price than when only Grade-A and linen were used. In other words, the synthetics are dominating the marketplace to the extent that both Grade-A and linen are becoming impossible to buy at *any* price.

In addition, we have another serious problem: ordinary cotton cloth is being sold as "aircraft" cloth, through recognized dealers, as Grade-A. It does *not* qualify in tensile strength or durability. There was some lame excuse like one special mill rinse to remove chemicals being eliminated! This *not*-Certified cotton does *not* carry the TSO—manufacturer's name, specs, etc. in the selvedge edge as required by the FAA. Why any aircraft supplier fails to notice this discrepency goes beyond understanding and into the realm of irresponsibility, if not worse. Both the mills providing this inferior fabric and the sellers are guilty of misrepresentation and false advertising. When a person orders Grade-A cotton for aircraft use, that is what he has every right to expect to be sold. Last fall we had a phone call from a friend in the Midwest wanting to know where we bought our Grade-A. They had been having this trouble. We had also heard of another instance out here in northern California. We offered a few suggestions, most of which they had already tried. Since our customers had been insisting on Ceconite, etc., we had no recent need to buy Grade-A. One of our suppliers is now handling Stits products exclusively, so we go farther for our covering materials now.

So *buyer beware!* Remember to *see* the required TSO marking in the selvedge before you buy Grade-A, and linen too—and *complain* if it is not there. It would certainly seem that both mills and suppliers are leaving themselves open to some serious lawsuits in this matter and perhaps one of the consumer groups could be of some service.

On the brighter side there have been some good developments regarding synthetic covering materials. There are now available pre-sewn cover envelopes (for all components) for nearly all the popular airplanes and these are most valuable time savers. Most fit pretty well but may require a little trimming and modification before being secured in place. Fit and pin in place, trim and re-sew as necessary as per Chapters 7 and 8. Be sure to try on each piece before securing it in place with cement. Prices for the presewn covers are quite reasonable in consideration of the time saved, and you do not pay for a lot of wasted fabric either.

Another nice product is the adhesive-backed rib reinforcing tape which goes on over the fabric, over the rib cap-strips. Rib-stitching goes over the reinforcing tapes. It sticks tightly in place during the rib-stitching and seems to be very durable. Also, rib-stitch cord and sewing thread is available in Dacron. The various TSO's for use of the synthetic coverings now call for all materials to be of the same basic chemical combination.

Pinked edge tape is now readily available in Ceconite, etc. This is a tremendous improvement over the early smooth selvedge tapes which never would stay doped down securely. This was a common failure on many planes just a few years ago. As we have long known, even the selvedge edges of Grade-A or linen are hard to cement down securely. In lieu of pinked edge synthetic tapes, pinked Grade-A tape served well for a long time over dacron cover.

Bias tape in Dacron (or Ceconite) is made now and it is generally used in three or four inch widths for going around wing tip bows and the round tips of the tail feathers, wherever the radius of turn is too great for the regular Dacron pinked tape to pull evenly and lay down smoothly. Standard Grade-A cotton tape can be pulled much better around a radius than can the Dacron.

This bias tape really stretches out and becomes relatively *narrower* as it is pulled, so be sure to start by first securing it at the widest portion and terminate at the narrowest or thinnest end of the run, such as from the leading edge around to the trailing edge of a wing tip bow. We have always done it this way with the Grade-A pinked tape, but with the new bias Dacron tape, it is imperative because it narrows so much as it is pulled. As with all tape, be very careful to keep it centered over the edge being taped. The bias tape is pretty expensive but is worth the cost for these few special places which do not amount to any great lengths. Use the standard pinked edge tape every place else where the full, uniform width is important.

Our current FAA Manual is now *43.13-1A*. Buy a copy and use it for reference regarding requirements. The diagrams in it are precise and easy to follow, especially for rib-stitching, for instance. Have this manual at hand and study it *before* you start to work.

Synthetics

It is absolutely necessary to follow the manufacturer's *precise* method of application, as per his STC if you expect to get your pride and joy approved and licensed. They have developed their methods for use with the specific materials. They are *not* compatible with each other except for the basic Dacron cloth. The FAA is insisting on following published methods of application for the good reason that there have been so many failures resulting from individuals "mixing and matching" according to their own misconceived ideas. In our own shop we strictly adhere to the established processes as published, such as the Ceconite Method or the Stits Poly Process, regardless of our preference for cotton or linen with butyrate dope. We follow the rules *for our own protection.* Should there be any failures with these materials, *"we only followed the published directions—explicitly."* We advise others to do the same. We cannot recommend experimenting unless a person wants to go into that business. I am not going to copy out the details of every approved method; they are readily available to anyone interested in comparing the various materials and there will be changes made in all of them sooner or later, possibly even before this goes to press. It would be superfluous, like someone wanting to know how to tie a rib-stitch knot—I look it up in *43.13-1A* every time I have to do it myself. You should have all that information at your finger-tips *before* you begin to work.

Because we have been using the Ceconite Method and materials more than others, I will go through with that in the following chapter. Here, very briefly, is an outline. It calls for two coats of silver over two of butyrate over two of nitrate; this is finished with just enough pigmented butyrate for a full and even finish color. We sand after the first coat of silver, quite thoroughly and extensively, because up to that point there has not been sufficient fill for it to be sanded effectively. Besides, it is then much easier to see where the sanding is really needed for a smooth and even surface. If this first cross-coat of silver is sanded off no harm is done; this sanding will be followed by the second cross-coat of silver which is essential for keeping out the sun's rays which are destructive to the fabric—yes, Dacron, just as much as cotton—and there is now a good base for the pigmented finish. Use 280 grit emery paper for this heavy sanding. Use 400 grit paper after the first color coat; at that time there should be very little sanding required.

Note: Before spraying, *every time,* use a cloth dampened with thinner to wipe dust, etc., from the surface to be sprayed.

Regarding masking: when using newspaper, double-thickness is necessary because wet dope will go through only one thickness. If butcher-paper is used, one thickness will suffice. Cover completely; 45 lb. force used for spraying will find any openings that are not *tightly* sealed with masking tape. When masking along the edges where two colors meet, a clean line between them can be assured by using a small brush to put clear dope along the edge of the tape (over first color). This seals that edge and the second color will not "bleed under" the edge of the tape, onto the first. Incidently, when removing masking tape, pull gently *away* from the newly done edge. To pull *toward* the fresh dope (or paint) would most likely result in having it come off with the masking tape. It is not advisable to leave masking tape over fresh dope (or paint) any longer than absolutely necessary; when removed, it may lift the dope or paint under it. This is particularly true if the temperature is high. Remove the tape as soon as the dope is dry to the touch. Best to allow about twenty four hours drying time before applying masking tape extensively.

Pre-Sewn Envelopes

Suppliers of pre-sewn envelopes are doing a good job and we do recommend them when synthetic covers are to be used. Home stitching of cotton or linen is not too difficult beyond the problem of bulk, but home machine stitching of a mass of Dacron could be a real mess. It can be done but it wouldn't be easy, for already described reasons, particularly because it is both heavy in weight and it is very slippery. A very large work table would be required, as well as a commercial type sewing machine. The accompanying series of photos in the next chapter illustrate the primary steps, from preparation of the airframe through color finish, using a pre-sewn fuselage envelope. The wings and tail feathers had been finished at an earlier date. This method can be compared with the earlier description of using linen with butyrate dope on the Fleet fuselage in Chapter 7. A well made envelope is a lot faster.

Except for the very small airplanes, there is no way to avoid the use of chord-wise seams in wing covers, whether the tips are sewn around the contour or not. There always remains the question of how best to attach the fabric at the trailing edge where there are ailerons and probably flaps. Some mechanical arrangement is usually provided to secure the fabric into these concave areas and this will depend on the method originally used. In some light planes, lacquer cement liberally used in the aileron well gets the job done. The Piper Pawnee, among many others, uses a strip of aluminum screwed in place to secure the fabric in the flap and aileron wells. Whatever is used must not interfere with aileron or flap action in any way.

One statement made in the first edition of this book must be corrected. I said that pinking shears could not be used to cut synthetic fabrics. *Wrong*. They cut much *better* than do straight scissors. So please accept my apologies—this is one reason books are revised. And again I emphasize the availability of Dacron pinked edge tape and also that the pre-sewn envelopes are now provided with all cut edges pinked. This material ravels something awful; when they had plain cut edges they were terrible and this was one big reason for my aversion to them.

Will your finish last "forever?" Absolutely not. Ask your FAA maintenance man. Forever is still a long way off.

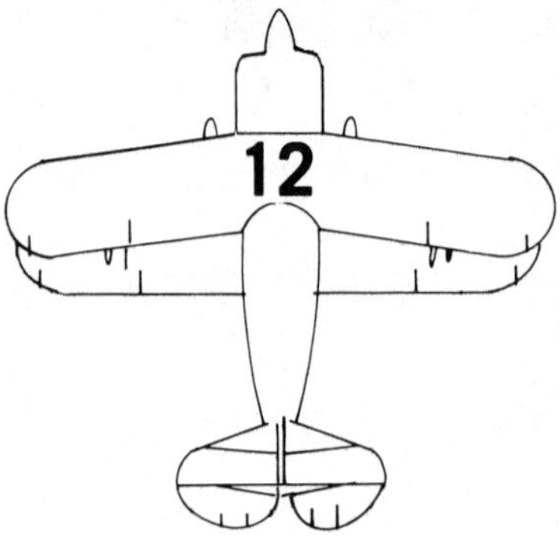

12

A Typical Recover With Ceconite and Pre-sewn Envelope

This particular aircraft is a 1941 Taylorcraft, BC 12-65 Deluxe model, and it happens to be one of a very few to have once upon a time had Fiberglass covering put on it—*over* the old fabric. I will digress to give a little background on this unsuccessful venture with which I had a brief experience (as a paid helper) some thirty-five years ago.

The objective was to develop a "permanent" aircraft cover which would be impervious to moisture, chemicals, etc. (Does this sound familiar?) Fiberglas cloth was new on the market and at first glance seemed to fit the need. But some very important factors were overlooked in the effort to sell the idea; as in too many cases, the *idea* was not thoroughly developed *before* trying to sell it. We see examples of this fault time and again. There were a lot of impractical features. There was not a good way to attach it to the airframe; the solution was to apply it over the old fabric or new muslin (if old fabric was gone), which was pre-stretched. It was very difficult to cut and it ravelled terribly. As already mentioned, there was no way to tauten the Fiberglas itself.

Now right at the beginning a most important matter was ignored: airframe pre-cover preparation. No responsible mechanic simply covers an airframe without a thorough inspection and repair as necessary. A Fiberglas resin was used to "dope" the Fiberglas to the basic fabric underneath following a sanding to remove the old, dry dope flaking from the surface. You can imagine the weight of this accumulation of old and new covers. Neither the Fiberglas nor the resin alone is light. Some kind of CAA approval was

Fig. 12-1. Sandblasted Taylorcraft fuselage ready for zinc chromate. This foreward area had the most rust.

arranged for T-Craft only, but just a few were ever covered in this way. There was also a Stinson being worked on at the time I was there but I do not know if it was ever completed. At any rate, this little T-Craft in our shop was in need of a lot of pre-cover attention although it had never been damaged in any way.

All of the steel airframe and fittings throughout were sandblasted. The most rust was in the cabin area where the wood framing of doors and windows had held moisture, and especially the landing gear legs inside the dirty, heavy, double cover where moisture had been trapped. In addition, insulating material had been put inside and around the entire front end—more weight! The present owner had bought the plane this way but had never flown it, which is probably just as well. He wants it right and has been very cooperative in making it that way.

Figure 12-1 shows the freshly sandblasted fuselage, ready to be sprayed with zinc chromate. We have learned that a fine spray of Aluminprep or similar dilute acid, as used on aluminum, will likewise help bond the chromate to steel. The tubular structure in the forward cabin area was painted black over the chromate and the instrument panel was also cleaned up, sanded with fine emery and painted with flat black; it was in good condition. All controls were re-installed with new cables, fairleads, etc., and control travel was checked at this time by temporarily mounting the surfaces. New wood window frames were made, but much of the old wood work (stringers and door frames) was sanded, revarnished and re-installed because it was basically sound. This Deluxe

Fig. 12-2. Instrument panel, all controls, wood formers around doors and windows and all stringers re-installed. Ready to be covered.

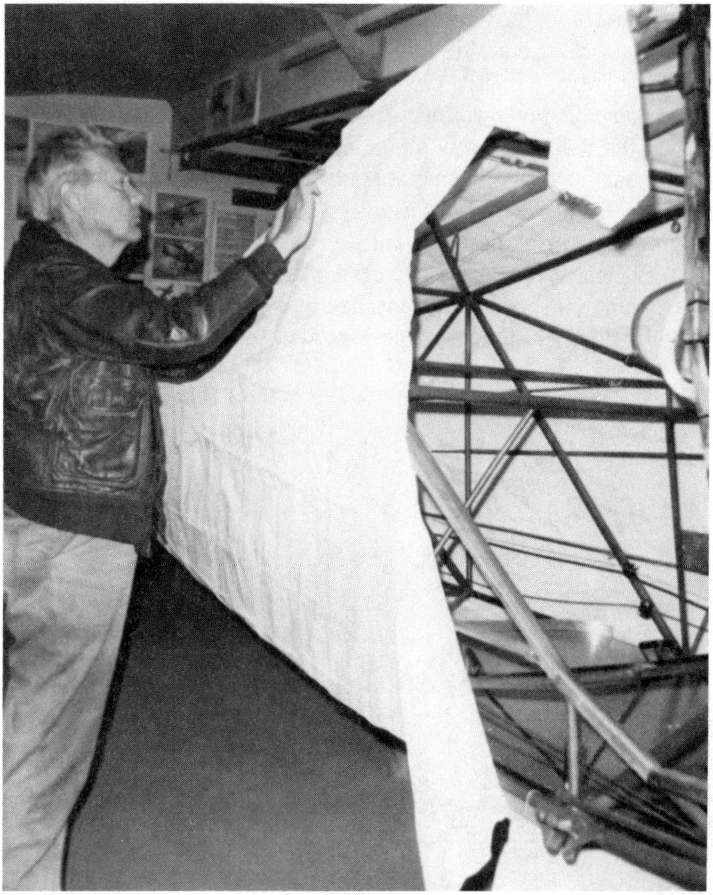

Fig. 12-3. Fitting the pre-sewn envelope at the front end after cutting out the door openings.

model came with fuselage stringers mounted in neat little metal brackets, giving the fuselage nice contours. Figure 12-2 shows the fuselage ready for cover.

Now we begin the fitting of the pre-sewn fuselage envelope (Fig. 12-3). Approximately six feet of the lower right rear seam had been left unstitched to allow the envelope to be pulled on over the fin which is mounted before covering begins. with the door openings cut out, a little trimming and adjusting finished it out quite well. The open seam at the rear end could have been stitched closed (on a fast airplane it must be) but we elected to cement it neatly to the lower right longeron, bottom edge up first, then the side edge down over that. Great care is taken to fit the envelope with the seams following the longerons, working the inside seam edges to lay along one side only of the tubing, for a neat and uniform finished edge, full length.

Figure 12-4 shows the Ceconite being cemented into place around the doors and windows after careful trimming for smooth fit. Excess material is being held temporarily in place to the side stringers by clothes pins (spring type). They are removed as the progressive shrinking takes up the excess. All edges must be secured before shrinking is begun. As shown in Fig. 12-5, all protrusions must have cut-outs made precisely and trimming is carefully done as the cloth is cemented to places like the fin post.

Fig. 12-4. Cementing fabric around doors and windows.

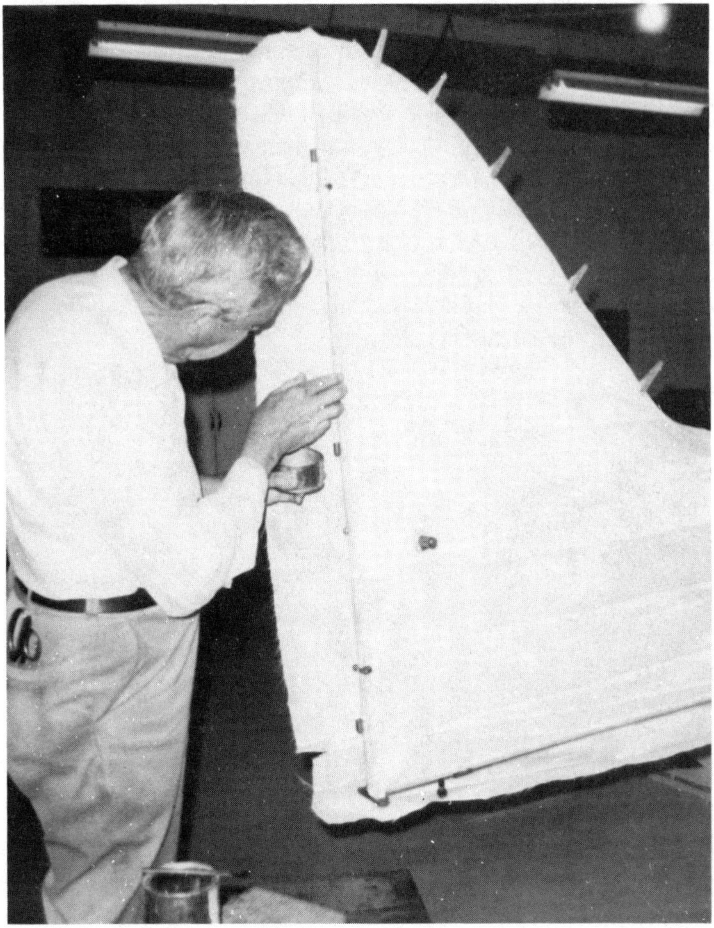

Fig. 12-5. Cementing fabric to fin post (right side) after trimming around hinges. Excess fabric of left side will be trimmed off before it is cemented over right side. Spring-type clothespins hold front seam of fabric straight along leading edge.

There can be no excessive thickness around hinges where there is limited clearance. Clothes pins hold the fin leading edge seam in place until the shrinking is done. Note that at no time is there any pulling of the cover; correctly made, there is no need.

With the fuselage Ceconite envelope completely cemented along all edges and openings, the iron is used (medium-low heat) to shrink the Dacron, working symmetrically to keep the shrinking uniform (Fig. 12-6). If one area, such as a top or side, were done too much at the beginning, it would pull unevenly from the surrounding areas and seams would be pulled all out of line and the cover would be twisted and unevenly shrunk. This requires a lot of patient work and watchfulness. If it is not evenly done there will be ripples in the completed surface. Compare this method of shrinking with the foolproof simplicity of just wetting well with water when using Grade-A or linen. After that, you go home for dinner and a night's sleep while it dries out and start doping the next day. You rub the water in and as it dries it shrinks by itself, perfectly.

When using either an iron or a heat lamp to shrink the Dacron, use great care *not* to hold iron or lamp in any local spots because that can cook the synthetic fibers! There would be a disaster. You don't want any patches at this stage.

Fig. 12-6. Shrinking Ceconite with iron (medium-low heat). Keep it moving slowly.

Fig. 12-7. Brushing on first coat of nitrate dope.

After shrinking to tauten the covering, the next step is to brush on the first coat of nitrate dope, being certain to get a full and uniform application with a minimum of back-and-forth brush strokes. Because synthetic fibers do not absorb dope as cotton and linen do, there is the unavoidable characteristic of the dope building up with repeated brush strokes; none of it soaks into the fibers. This is why the finish is prone to peel off these materials. In Fig. 12-7, showing the initial brush coat of nitrate being applied, the dope appears to be white, in contrast to the dark appearance of the first coat of dope being put on linen in Fig. 7-18. Because of this white appearance, it is very difficult to detect "blush" of the clear dope; it has an opaque, rather dull look, as compared to the clear-white it should have. If the atmosphere is dry and warm there should be no problem. Also, in comparison with the first coats of dope on cotton or linen which result in considerable sagging of the water-tautened cloth, the Dacron does not sag appreciably ; it's like a cotton or linen after the second or third coat when it started to tighten up again. It *does* tauten progressively with added coats of dope, but not very much. Unless it has been shrunk too much with the iron or lamp, the few coats (total) of dope to be used will not cause it to over-tighten. Often, when an airplane sits out overnight in cool damp air, the fabric will look saggy in the morning. Nothing to worry about; it will become taut again as it dries and warms.

Fig. 12-8. Applying the pinked tapes.

Fig. 12-9. Fuselage is ready to be masked for silver butyrate. "Boot" cowl goes on forward of door. Interior will be finished by professional upholsterer, using a soft grey material.

Following the two brushed coats of nitrate dope, the pinked tapes are put on as described in Chapters 7 and 8, as done with cotton or linen (Fig. 12-8). Do not pull over hollows or irregularities, such as between the rib-stitches of the wing and tail surfaces.— Yes, the rib-stiching is done after the two coats of nitrate *before* the tapes are put on, just as with cotton or linen. The only place that tapes are pulled is where carrying them around wing tip bows and rounded tips of tail surfaces; when working with the synthetics, bias tape is used there as previously explained. Butyrate dope will *not* hold the dacron fabrics together nor to anything else. It is used for fill and for the finish coats *after* all tapes and reinforcing patches, etc., have been securely doped down with nitrate. And we definitely like buty on the surface because of its far greater flame resistance, though if it starts to burn, it is going to go fast with that nitrate base.

A helper is needed to put pinked tapes over the stringers of a long fuselage. To get the tape on while the dope is wet, only a couple feet or so can be done at a time, so a second person can help by watching to see that the long tapes are going on straight, and also see that they stay centered over longerons, surface edges, etc. As a section of tape is put on, immediately brush over it with

Fig. 12-10. Spraying the silver butyrate. The top has been done first.

Fig. 12-11. Note wrinkling on fin. See text for cause and cure.

more dope and make certain that it is *completely* stuck down so that there are no bubbles nor lifted edges. It is necessary to build up the dope fill of the tapes to match that of the base fabric so the total will be uniform for a good finish. Lay out and cut the pinked tapes to fit neatly at corners and curves, without overlaps. Sloppy matching of cut edges, etc., indicates poor workmanship.

We now look back into the fuselage which is ready to be sprayed with silver butyrate dope (Fig. 12-9). The tapes have been doped and lightly sanded, and the whole thing looked over critically for omissions. This is the time for any corrections or additions.

The silver is now sprayed. Figure 12-10 shows a nice fan from the spray gun, with about 45 - 50 lbs. at the regulator. The adjustment of the spray gun for volume of liquid output cannot be positively described. It must be shot on a large piece of cardboard or heavy paper and adjusted by trial and error to determine the best setting for the mix of dope being used. We have found a great difference in various manufacturers' dopes. If it goes on too dry it will have a sandy look and feel, like overspray (it amounts to the same thing). If too thin, wet, or too great a volume, it will run, puddle and be full of air bubbles. We are always amazed at all the fine bubbles in the freshly sprayed dope on any of the Dacrons, as

Fig. 12-12. The first coat of white over the silver, to be followed by two additional full cross-coats for a nice uniform white. Yellow on the lower third was sprayed later when white was dry enough to be masked. A light green stripe will be added. Wing tips are also yellow.

compared with the spraying on cotton or linen. It probably comes from the greater air spaces in the wider-spaced weave of the synthetics. At any rate, they eventually fill and the bubbles disappear as the dope dries, so don't get excited about it. When it is dry enough to sand, they won't be there. In the photo, the top of the fuselage has been done first; the bottom will be done last because that results in the least amount of over-spray. Remember that a full cross-coat consists of spraying vertically all over and then going over it again with fore and aft swaths. This gives uniform coverage with no striping. This system is followed on every fabric covered piece, including landing gear V's if they are covered.

After the first full coat of silver, use 180 grit emery paper for a very thorough sanding down. Don't worry about removing the silver; the second cross-coat will put a sufficient amount back on, according to the method now in use with Ceconite. The very few coats of dope with Dacron is the biggest difference between that and natural fiber coverings. Here, the number of coats is kept to a minimum; keep it thin so it won't crack. Until the silver is on, you really cannot see well where the sanding needs to be done, nor is there sufficient build-up to be sanded effectively. so this is the time to sand out excess dope (the silver shows up the brush strokes very plainly), remove any blemishes, and provide a fine smooth surface for the color which will follow the second cross-coat of butyrate. The second coat of silver builds up the base for color and it is important for reflecting the damaging rays of the sun from the Dacron, which does not last a whole lot longer than natural fibers when exposed to the sun. The color of your choice is primarily for appearance.

Figure 12-11 shows finishing the first coat of silver at the tail end of the fuselage and it also shows something unexpected which provided a lesson for us. Note the wrinkles above the access hole (same on both sides). The factory had provided thin plastic, L-shaped frames to reinforce the holes; metal plates screw to them with P-K screws. The fabric had been neatly cut, folded in and cemented inside with lacquer cement and everything looked fine. But as the sprayed silver dope softened the nitrate on the fin, the Dacron tautened upward as it dried and pulled up the fabric around the hole where there was nothing much to resist. If left alone, there was going to be a very distorted hole. The cure was to drill the holes for the screws, run lightweight wire through the holes and around the stabilizer rear mounting tube. The wires were gently tightened while the doped Ceconite was still soft and the access-hole frame was pulled straight. The wires were left in place until all doping (including color) was completed. The other way to go would be to wait until all doping was finished before cutting the hole and then fold in the hole edges and cement to the inside of the plastic frames. However, this would risk making a mess of the finished surface around the holes, nor is it easy to fold in and cement edges of cloth that has had numerous coats of dope on it first. So if you can't avoid a rather large opening, pre-wire the edges together or to something solid within reach.

Figure 12-12 shows the white butyrate dope being sprayed. For a true, uniform white over the silver, we used three cross-coats, sanding lightly with 400 grit emery paper all over after the first full coat was completely dried. We caught a few minor runs; this is the time to sand them out. The last two coats of white were sprayed on very carefully, and lighter in volume.

Remember, when spraying dope—or painting anything, for that matter—from the clear dope through the finish color, first do all the edges, pockets and crannies which would otherwise be passed over too lightly. These places are done *first* so the surface spraying will blend it all in nicely. Any local spots sprayed after the full surface always show up the additional thickness of dope; in addition, surrounding over-spray is inevitable.

The lower third of the fuselage was sprayed with Lemon Yellow. The wings have yellow tips and there will be a light green striping to complete and complement the white and yellow. Two cross-coats of the yellow-over-white were sufficient for good depth of color. Total masking of the fuselage is necessary so no color gets on the white where it is not wanted.

The interior of the cabin is being re-upholstered with a soft grey material. All metal parts will be painted with matching enamel. This will require another full masking so trim-lines will be correctly matched. It is the masking that takes the time!

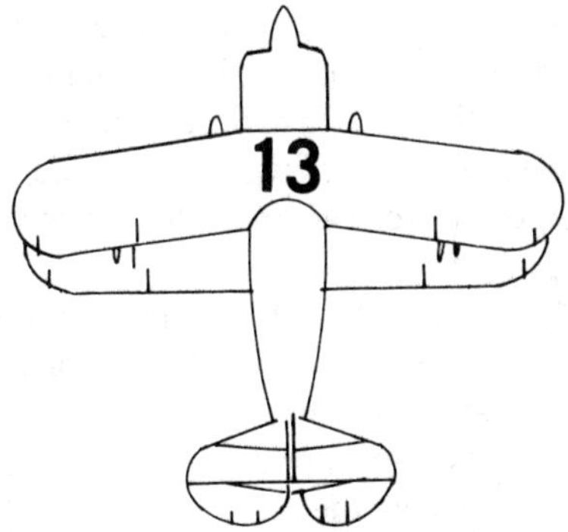

Appearance: Yes—Protection: Yes, Indeed!

We get an increasing number of requests to "fix" cracking finishes on fabric-over-plywood skins of relatively new production aircraft, although some are getting older than you would think. There is no way to accomplish a good quick fix. Time is money and skinned aircraft are loads and loads of time and costly plywood. The way to do it *right* is to completely strip off all of the old finish and start over. Yes, small cracks and chipped out abrasions can be sanded and scraped down to bare fabric, but the primary concern with these problems is the trouble that has developed *underneath* in the plywood skin or even under that. This is the reason for certain AD's on some of these airplanes.

Moisture and dirt work through the cracks and holes in the finish and into the wood. Being closed in, the spreading moisture cannot ventilate and dry out. The plywood will absorb the trapped moisture and warp, losing its original contour, and it gradually loosens from the structure to which it was secured by glue. Of course this results in loss of strength of the component. If this becomes excessive, a whole section of wing, or other surface, will require major repair and replacement of bad wood, or if less serious, by at least sanding and re-gluing if there is no deeper deterioration present.

Wherever any woodwork has been done, sand and re-varnish all *interior* surfaces involved. (See *Aircraft Woodwork* by Ruth Spencer, TAB book No. 2204.) Use nitrate dope as sealer over new, or sanded old wood, where a Dacron fabric will be used to recover the plywood skin. Use butyrate dope if the fabric is cotton or linen. Apply cover with a two inch overlap of old cover, cementing it down with lacquer cement all around the edges. When dry, shrink with an iron or lamp, taking care not to have too much heat with the plywood under the cloth. Proceed with brushing on the nitrate, sanding, etc., as with any other covering, including pinked tapes.

It is apparently common practice to surface plywood skin with filler at the factory before fabric is applied. When there is surface cracking and chipping over these filled areas, the wood deterioration is even greater, because these filled areas were to level hollows in the first place. The filler only helps to retain the moisture, making a bad situation worse. Nor do we know what effect the dope will have in combination with some fillers. In any event, fillers should be used very sparingly because of the weight factor.

With any part of an airplane the biggest problem areas are usually along or near the trailing edges where drainage and ventilation is so often poor. Retractable landing gear wells are particularly vulnerable to loosening glue joints. They must be kept clean and dried out. It is advisable to scrape and sand away dirt that accumulates and repaint the insides of the wells periodically, paying special attention to the many nooks and crannies. Most aircraft owner's handbooks give advice regarding this subject, from the standpoint of the care of wheels and brakes, if not the wing structure.

Do not overlook the condition of flap and aileron hinges nd the bolts and bushings used for their attachment. Here again, these are all near the trailing edges and particularly subject to rust or corrosion. Use the correct sizes of replacement bolts, nuts, bushings and washers—using wide-area washers against wood. Over-length bolts can impede proper control action. Castle-nuts with cotter-keys are still advisable. Use self-locking nuts at hinges only if specified by the manufacturer, and where these are used, put washers under both bolt head and nut. *Never* over-tighten any bolts through wood structure. That crushes the wood fibers and results in strength loss. Rusty or corroded bolts and bushings are that way because of moisture, so if you find them in that condition, the surrounding wood is suspect.

Wing Root Rot

The most serious example I have found in a production plane, less than ten years old, was rot at the right wing root under the wing-walk of a Bellanca Viking. The history of the aircraft indicated service in a South American country of high humidity—and definitely a lack of maintenance overall. Repairs to the plane had already been made, including work I had done on the left wing, following a landing accident. An observant aircraft painter spotted a questionable appearance near the right wing root and leading edge when sanding it down in preparation for complete painting. Enamel was to be put on over the butyrate finish, when the entire airplane *needed* a complete recover. To be brief, the fabric was removed from the wing-walk area; the old non-slip material over the walk was badly deteriorated and flaking off along with the fabric covering beneath it. When this was all removed, the worst part was *not* forward, as thought (where the fuel covering piece of plywood is held by screws, one inch apart all around), but to the top surface of the root at the rear spar, with a vague line wandering outboard toward the aileron well. See Fig. 13-1.

I expected to do a skin graft, replacing the plywood over the affected portion, but as the plywood was removed from the top surface of the spar (rear), the condition became worse approaching the butt end where the wing attach fittings are bolted on, a pair front and rear at top and bottom. It was soggy wet there, soft rot down to over an inch deep. The top inboard attach bolt goes through there just a half inch below the top surface of the spar (Fig. 13-2). My excavation with a wood chisel went well below the bolt, so right then I said, "No," considering the Bellanca wing construction. The nearest Bellanca service agency says there is an approved fix, but it would involve more time and work than practical in my opinion. A good replacement wing was advised—the high cost of neglect! This had begun as a dope and fabric condition. The structure aft of the spar was sound. It looks rough in the photo because all of the plywood had not been removed to bare rib structure.

We are glad to see the makers of synthetic coverings calling for re-cover in something less than a "lifetime." Any mechanic and all owners should know that a machine needs a lot of work inside after a period of years. Incidently, in regard to the long time linen covering on Spencer's Stearman fuselage, I must say that *many* times we have been down into the back end of it, via the baggage compartment. This is possible with a Stearman, but not many other planes. It is possible that the Dacron covers *may* hold up better than linen; only time will tell. But as long as there is trouble with cracking finishes, the most definitely will not. Outside tie-down is an unhappy fact of life in most parts of the country now.

Stits Poly Fiber

Little has been said here about the Stits Poly Fiber Process, partly bcause we have not used it much. It is more expensive and we do not think it is that much better than the other synthetics. One or two jobs were done through silver, by the Stits process, then we turned the enamel finish over to the owners. We have seen some very nice amateur coverings using the Poly Process; we will watch them age with an open mind. There has been a lot of improvement in finishing all of the synthetics during the last few years. People have learned that the only way to make things stay together is to

Fig. 13-1. Fabric removed from Bellanca right wing root.

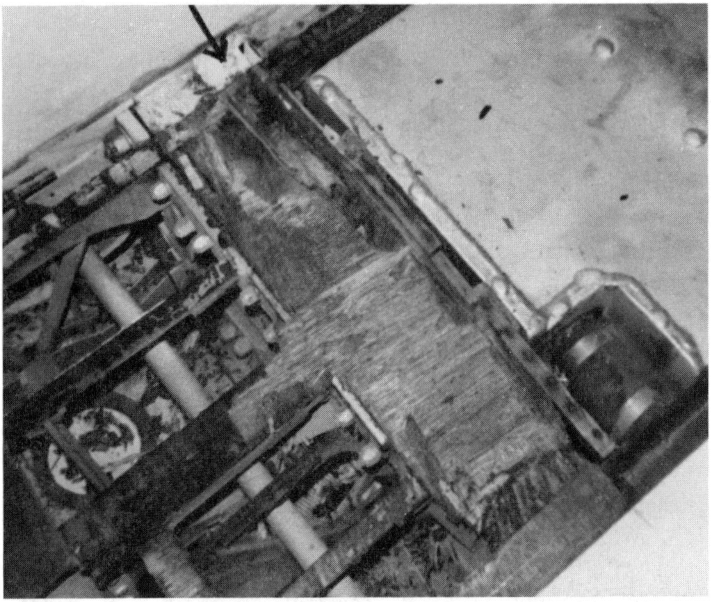

Fig. 13-2. The deterioration exposed when plywood skin was removed. Lack of proper maintenance does not pay.

keep the finishes thin and light; "beautiful" many-coated, hand-rubbed finishes are *out* for synthetic covers because they do not last; there is nothing beautiful about cracked and chipped finishes. It is now accepted that the weave of the fabric *should* show in Dacron covers. With 80 threads to the inch in Grade-A compared to 53 per inch for Dacron, you can draw your own conclusions regarding the comparative finishes possible. Both Grade-A cotton and linen absorb and hold a lot more dope for a much finer finish. Finer weaves in Dacron, as in cotton, may be used on ultra-light aircraft, but our primary concern is with more conventional, open-structure airplanes requiring more tensile strength in the fabric. Anyway, we see nothing objectionable to seeing the weave in fabric. If a person wants an airplane to *look like metal*, he should buy a metal airplane. They can have deep enamel or lacquer finishes with no harm done. The best enamel-on-linen job we ever saw was on a friend's airplane. The black enamel had been very lightly sprayed over old butyrate dope. Black covers very well so it took only a light (in volume) coating for an excellent finish. Because it was light and flexible, it lasted well for a number of years until the fabric was due for replacement.

A Parting Word

Now that you have reached the end of the book, if you are really planning to cover one (or more) yourself, I advise you to read it all over again. In studying anything, you don't just read through a book and expect to retain all you have read the first time through; you *study* it. You try it out on samples and gradually what was written will begin to come clear in a practical way so that you can *use* it in an understanding, knowledgable manner. This is certainly true in working with aircraft. Every item in an airplane is there for a specific purpose; it is not to be changed without good cause by someone who knows what he is doing. The weight factor is never to be overlooked; excessive weight can cost dearly in critical situations and so can altered wing contours which result in loss of lift. There is much to consider besides a pretty finish. The most important thing is the aircraft structure and its proper maintenance. A good surface helps maintain the structure beneath it.

Whatever the materials used, it is the *workmanship* that makes the difference. There must be care taken with all details. Do not get impatient. That does not mean to work too slow. Working with these materials which dry so quickly, you have to move right along. Each step must be planned ahead with an understanding of precisely what is to be done. Patience brings its own reward; one's pride in a good job is well deserved and your airplane will take better care of you.

I admit to impatience when a person tells me he "has read the book" but he wants to ask some questions, then the questions indicate that he has not "really" read the book at all. What he wants is to be taken by the hand and lead along, step by step. There is no magic easy way to do this kind of work, it is *work*. My pleasure comes when a stranger approaches at a Fly-In to say he wants to thank us for the book which was really a help to him. Good.

Our personal interest is in the restoration of old airplanes and construction of new ones, especially of the sport types such as the pair we own and fly. New production aircraft are built for fast transportation, not for the joy of flying. The satisfaction of flying a plane which you have built or worked on is in a class by itself. Yes, we do this work for our living, but at every opportunity we encourage *qualified* people to do their own if they have a licensed A & P Mechanic to *supervise* their work closely. The purpose of the book is to assist those new to covering airplanes, whether amateurs or professionals. *Every* job should look professional, not "good for a beginner."

Editor's Note: At present, there are four synthetic fabric/finishing systems approved by the FAA for installation on ATC'd aircraft: Stits Poly-Fiber, Ceconite, Eonnex, and Razorback.

Razorback is a pre-treated fiberglass cloth finished out with butyrate dope. Stits is a square weave Dacron finished with resin compounds especially devleoped for use on Dacron. Ceconite is a Dacron cloth finished with nitrate and butyrate dope. Eonnex is a pre-treated Dacron that is finished with only three coats of specially developed water-based emulsions.

The "no-name" Dacron cloth offered by aviation supply houses is not a part of an FAA-approved system and may not legally be installed on ATC'd airplanes.

Index

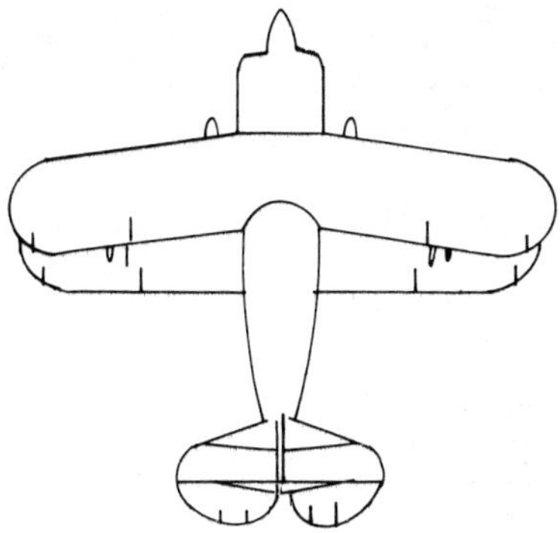

Index

A

AAA 14
AC 43.13-1 15
Aileron hinges 148
Air compressor 106
Airworthiness Standards 118
Aluminprep 134
Amateur builder 16
Anti-drag wires 36
Anti-tear strips 93

B

Bach, Richard 112
Base color 116
Bias tape 128
Blanket covering 84
Bleeding 116
Blushing 118
Bug sprays, oil-based 111

C

Canvas 15
Ceconite method 129
Celluloid 11
Center-section 99
Color schemes 115
Color sketches 115
Contour measurement 55
Cotton, inferior 127
 TSO 127
Covering, life of 124
Curves 117

D

Dacron 123
 shrinking 137
 thread 128
da Vinci, Leonardo 9
de Havilland, Geoffrey 11
Dope brush 71
Doping 69
Drag wires 36
Drainage 31

E

EAA 14
Earth colors 114
Enamel, and butyrate 104
 patching 102
Envelope, seam-sewn 55
Eonnex 153

F

Fan pattern 107
Fiberglass 132
Finish removal 125

Flammability 18
Flap hinges 148
Fuel tank inlets 99
Fungicide 51

G

Glue, removing 40
wood 36, 44
Grade-A cotton 19
Grommets, drainage 97

H

Hall, Ben 23
Headliner 50
Headrest, aluminum 102
Humidity 23

I

Insects 111
Inspection holes 79
Interior covering 48
Interior trim 53
Irish linen 123
Ironing 61

K

Knots 95

L

Landing gear wells 148
Lifetime covering 149
Lift struts 101
Lights, covering around 96
Linen 77
life of 149

M

Manual 18 15
Masking 116
stencils 117
tape 117
Measuring 55
Metalizing 27
Moisture accumulation 101

N

Nails 36
Newspaper as masking 117
N numbers 116

O

Overspraying red 125

P

Paint stripper 125
Patches, reinforcing 75
Patching 102
Planning 55
Pliofilm 23
Plywood 147
skinning 99
Poberezny, Paul 13
Pre-sewn envelopes 130

Q

Quick fix 147

R

Racing aircraft, covering 93
Razorback 153
Red dope, overspraying 125
Re-doping 102
Regulator 106
Rejuvenating 124
Retarder, dope 70
Rib repair, aluminum 30
Rib stitching 93
Rot, wing root 148
Rust 28

S

Sandblasting 28
Sewing 60
Silk, clear doped 11
Silver dope 105
Silver paste 107
Shopping lists 21
Shrinking 65
Slanted numbers 117
Spar ends 83
Spray gun 107
Stits Process 126
Streamlining 101
Surplus dope 22
Synthetics 129

T

Taping 72
Taylor, Robert 13
Temperature 23
Trim color 116
masking 116
Traditional materials 17

V

Varnish 41
Ventilation 23
Visibility 114

W

Water shrinking 65
Water trap 106

Wax 20
Weldwood 44
Wing rigging 100
Wing root rot 148
Wooden wings 34
Workmanship 152

Z

Zinc chromate 134